MW00896856

THE ULTIMATE EASY TO HARD SUDOKU CHALLENGE

INTRODUCTION

Welcome to "The Easy to Hard Sudoku Challenge". In this collection of puzzles, we celebrate the timeless joy and enduring benefits of Sudoku.

Sudoku, with its blend of logic, strategy, and problem-solving, has emerged as a beloved pastime around the world.

Beyond providing a delightful way to pass the time, studies have shown that engaging in activities like Sudoku can enhance memory, improve concentration, and even potentially reduce the risk of cognitive decline.

So, whether you're a seasoned Sudoku enthusiast or just discovering the joy of this timeless puzzle game, dive into this collection and experience the thrill of solving Sudoku puzzles tailored to people of all skill levels.

Get ready to sharpen your mind, challenge your intellect, and embark on an unforgettable Sudoku journey!

INSTRUCTIONS

Fill in the grid so that every row, every column, and every 3x3 sub-grid (also known as a box or region) contains the numbers 1 through 9. Each number can only appear once in each row, column, and 3x3 sub-grid.

- Begin with an empty 9x9 grid.
- Some cells already have numbers filled in as clues.
- Start by examining the filled-in numbers and the constraints provided by the rows, columns, and 3x3 subgrids.
- Deduce the possible numbers for the empty cells based on the given numbers and the rules of Sudoku.
- Continue to fill in numbers until the entire grid is complete, following the Sudoku rules.

TABLE OF CONTENTS

EASY - 1

	3				2	9		8
2							6	4
	6	7	8			2	3	1
1	2	6	7	8	9			5
3		5		2		7		
		8	5	4	3	1		
	5	4	2	7	1		9	3
6	1	2	9	3	8	4	5	7
9	7			6				

EASY - 2

	1		6	3	4	7	5	2
7		5	1	9		4		3
3			7		5	8		1
6	9		8			3		
5	4		3		7			9
	7				9	6		5
2	8		5	1		9		7
4		7	9		2			6
9	5		4		6	2	3	

EASY - 3

5	6		9			8	3	
3	2		6	4				
		8			2	4		6
8	3				5	1	4	7
			1	7	6			8
	7	2	4		3	5	6	9
	4		5	2		9	8	
	5	3	8	6	4		7	1
		1	7		9	6	5	4

EASY - 4

8		2	7		1	4	5	9
	7				5	6	1	
1	5	9	4		2	8		
		7		5	9		8	
5				1	4	2	7	
6	1	8	3		7	9		
7		6	1	9		5	2	4
4			5		6	3		
9	8	5			3			1

EASY - 5

3	1	6		7	4		9	8
	4	2	5		3	7		6
	7	5	6	9	1			2
	6		3	5			7	
	3	9	1		7	8		4
	8			6	9		5	
	9		8		5	6	2	3
4	2		7				8	
	5	8			2	1		7

EASY - 6

3		1			8			6
	5	6			7	4	3	8
4	7	8		6				5
5				8			1	3
8	3	2		9	1			7
	1	7	3			9		
2	4			1	6	3	7	9
		9	4	3	5	2	6	
1	6		2		9		5	4

EASY - 7

8					3			
5		3					9	7
7	1	6	9	5		8		4
6	9	2	1	4				
1	5			3	9	4	6	8
		4	6	7	5			1
2	3	5	4	9	7	1	8	6
4	6	8	5	2	1			
					6	2	4	

EASY - 8

2	5	9	4	6			1	7
7	3	8	9		1			2
	6		7		3		5	8
				8	5	1		
	1		2	9	7	5	3	6
6	7	5	1	3	4	2	8	9
	4	3	8		2		9	
5		6				7	2	
9			5					3

EASY - 9

		4	5	1	2		8	6
		6		3	7	4		
8	7	5	4	6			2	3
7	9	2				5	3	8
3	4	1	9	8		2	6	7
6		8	2		3			4
4	8	7		2	1			9
2		9		5	8		4	
	1					8	7	

EASY - 10

2	8		6		4			
	1	4	9	3			5	2
6	9		2	1	7		3	4
	2			4	5		9	
3	7		8	9	6		4	
5			3		1	7	8	6
	6		5	8	9	3	2	
9				7			6	
1	3		4		2		7	9

EASY - 11

7	9			8		1	2	
	5		9	2	1	7		3
		1	7	5				
	8	4	1	9	7	6		2
9	6	7		4	2			1
1			8		6	4		7
5		6		1	9	8	3	
	3	9				2	1	6
4		2			8		7	9

EASY - 12

3	1	6	2		5			9
5	8			1	4		6	2
2	4	7		6	8		3	
8	6							5
		5	8	9	1		4	6
	3			5	2		1	
4		2	1	8			5	3
	9		5	4	6		7	8
		8		2	3	1		4

EASY - 13

7	4	5			3	6		9
3				5	7		4	1
1	6	8	2	9	4			3
6	7		5	2	8	9	3	4
8		2		3		7	1	6
			7	6				
5	1	6				4		7
2			9		6			5
	8		1	4		3		2

EASY - 14

	9	2	5	8				7
5	8		3		4			
3			1	2			5	9
	1	5					7	
9		4		7	1		6	2
				5	2		1	3
7	5	9		4	3	6	8	
4	2	1	9	6	8	7	3	5
8		3		1	5		9	

EASY - 15

9	2	1	6				5	8
	5	6			8	3	7	
3			1	4			6	9
5			2		6		8	3
1	3			8	9	7		
6	8		4	1	3	9	2	5
7		4	8		1			2
2		5	3	9	4	8		
			7	5				4

EASY - 16

6				8	7	4	1	3
1		8	9			6	2	
	2	3	1				8	9
5	6		4	7			9	
9	3	4	6		1			
8			3		9	2	6	4
			8	1		9		
3		1	7	9	2	5	4	6
2	9	6		4				8

EASY - 17

5		2	4	6		3	7	
9	6	3				8		4
			1		9	6	5	2
		8			5	1	4	
6	7	5	9		4	2	3	
		1	6	8	3	7	9	5
8	3			7	6	4		1
4		7				5	6	
1				4	2	9		

EASY - 18

6		3			2	5	4	1
			6		4	8	3	
	4	1	3		7		9	2
5	2	6	1	4	8	3		
	8	4	5	3	9			
3	1		2	7	6	4	5	8
			4			7		3
4		7	8	9	3		2	
		5	7		1			

EASY - 19

2				4			1	9
1	8	4	3			2	7	5
9	7					3	4	6
5	1	7	9	2	3		8	4
8	4	6			7			2
3	2	9		8	4		5	1
	3				1	4	6	
	5	8		7	2			
		1		3	6		2	

EASY - 20

1	4		9			3	6	2
	3	9	4	6			8	
6	7	5			2		9	
5			6			1		
			2	5	4	8	7	3
4						6	5	
3	5	1	7	8	9	2		
8	9		1	2				7
7	6	2	5	4	3	9	1	8

EASY - 21

3			6	7	9	4		8
4	8	9	1	2	5	7	3	6
					8	9	5	1
5		4	8	9		6		
7					6		9	4
8	9		7	4	2	3	1	5
2	6	5			7		4	9
9	4	7	2	6		5	8	
1	3		9		4			

EASY - 22

6	4	9	8	5	7		3	
1	2		9				7	8
7	8	5					9	4
3		8	5	9	2	4		
2	1	7	3					5
9	5	4		7	6		2	
		2	6		9		5	
		6	4	1	3	2	8	
8					5	3	4	

EASY - 23

				4	5	2	7	
2	4	5		7	8	1	3	6
		3	6	2		4		9
4	1					8	9	7
	2		7					
5	3			8				4
	9	4	5	1	7	3		2
8		1	4	3	2	9	6	
3	5	2				7	4	1

EASY - 24

9	5	2	6	4	1			3
4		3				5	6	1
1	8	6	3				9	4
2		4	8	7	5	9	3	
5	6	7		3	9		4	8
3	9	8				7		5
						4		
	4	9	5	2	8	3		
		5		1	3	6		9

EASY - 25

8	2	6	9		4			7
9				7				
	1	5		6	2	8		
2			1	9	6	4		3
				2	3	6	9	8
3	6	9	7	4		2	5	1
1		2				9		
5	8	3	4		9	7	2	6
		7		8	5		1	4

EASY - 26

2				4	7	8		
6	9	4	5	8	3	1	7	2
7	8	3	6					5
	7		3	5	9	2	4	
3	5	2		7		9		
8	4	9	2	1	6	3	5	7
9	2	7						4
4			7	6	8		2	
						7		1

EASY - 27

	7	4	2		1	6	9	3
2	6	9			7			
8	3	1	9		6		2	7
7	4	3				2	8	
	9			2	8	3	6	
		8	3		5			
			1	7	2	9		8
9	1	7			3	5	4	2
3	8			9		7		6

EASY - 28

			1	7		4		3
7	4	9		6		2		1
			2	4			6	5
				2	4	3	1	
1	7		8	9	6	5		
		2	5	3	1	6	7	8
2	1	7	9	5			4	
3	8			1	2			7
9		4	6	8			3	2

EASY - 29

	2		6		1	9	3	5
3		6	2			1		
		5	9	4	3	6	2	
9	3	2	7		8			
	6	7	5		9	3		1
5	8	1	3			7		2
7			1	9			6	3
2	5	9	4				1	
		3	8		7		4	9

EASY - 30

3	8	7	2	6	1	9	4	5
9		4	7		5	2		
2			4			1	3	7
		1						9
6		5	1	2			7	4
	2	3	8	9	4		5	
		2	3	1	7	4		
4	3	9		5		7	1	2
1			9	4			6	

EASY - 31

6	2	8	5	9	7	1		3
9	4		6		3	8		2
5				4	2		9	6
	3	6	1			4	7	
8	7		3	2	4	5		9
			7	8	6	3		1
	8	4	2		1	9		
1	6	9	4	3		2		
	5			7	8			4

EASY - 32

9							8	4
4		8	7		6	9	2	3
3	2	5	9	8				6
5	9	6		7		3		
		3	1	4	5	6		2
	4	1	6		9			7
1	3	2	8	9	7	4	6	5
7	8		5	6				
	5				1	7		

EASY - 33

7	4		5	6		1	8	
5	9	3		8				7
	8	1	2	9				5
		7	8	3		6		1
	3	4		1		5	7	2
	6	5	7	2			9	
			6		9	7	2	4
	7	9	1	4	8	3		
4		6	3	7		9		8

EASY - 34

	3		8			9	1	7
4					7	8		
7	8		6	9	1	5	4	3
		4	5	3	9	6	8	1
	5		4		6		7	
6				2		4		9
	2		9			7		4
8	6	9	2	7	4		3	
3	4	7	1	6		2		

EASY - 35

		5		2	9	6		8
9	2	6	4		5			1
4	7			3	1		9	2
2	9	1		4	6		7	
7		3	1					5
	5	4		7	8		2	9
1				5	3	9	8	
		2	9		4		5	7
	3		8	6	7		1	

EASY - 36

6	5	2		8	9			3
9	3	8	5			2	6	7
7		1		3	2			5
	8	4	1	7		3		
2	9	6		4	3	7	5	
3	1	7	9	2	5			6
1		3		9		5	7	4
4					1	6		
					7			9

EASY - 37

2	8			6	9		5	3
9	3		8		5			2
5			3		2	8		
			1	2			4	8
4	5	3	9		7			6
	2	8			6	3	7	
8	4		7	9		6	3	5
3	9		6		8		2	7
	7			3	4	9		1

EASY - 38

8	9		5	4				6
	5	4	9	7		1		8
1			3	6		9	4	
	1		6		4	8	7	3
		8		2	5	4		9
4	6	9	8	3	7	5	1	2
	4	2	7	8		6		
7		1				3	9	4
	3				9			

EASY - 39

7		5				6	4	
	4			7	1	2	5	3
	1	6		5	3		9	
8	7		1			9	3	6
6	2		7	4		1		5
	5	9		6		7	2	
5	9		2			4	6	
		7	9	8		5		2
	8		5		6	3	7	9

EASY - 40

	8		7			1	2	
		5	1	2	9		8	4
	9	1	4		8	6	5	
9	2	8				7		
5			6	7	3		9	
	3	7		9	2	5	4	1
	5				1	2	7	6
			9	8				5
	4	6	2	5	7	9	1	8

EASY - 41

2	4		6	8			5	9
7	3	8	9			4		6
	6	9	3	2	4		1	7
	8			6		1		
6	2	3	7		1		8	
1	7		5					
			8	7	9	6	4	
4	9		1	5	2		3	
8	1		4		6	5		2

EASY - 42

	6	3		9	7		1	
4		9			1	2		
8	2	1		4	5		3	7
			4	3	6	7		
	1	6	9	5				4
	4	5			8		9	
	8		5	2	9	1	4	3
9	3	2	8				7	6
1	5	4		6		8	2	

EASY - 43

7		3		1			4	8
9	2		7		4	5	1	3
				2	5	6		9
	8	2	1		6			
	9		2	4				
			9			4	6	2
8		1			2	3	9	4
5	3		8		1	7	2	6
2	6	9		7	3	1	8	5

EASY - 44

		7			4	2		8
9		4		1	2	6	5	7
2			5	7		4	9	3
	7	9	2	3		5	6	4
1	4			5				
5		3	6		7			
3		8	1	2	9	7		
	1		7	6	5	8	3	
7			4	8			2	6

EASY - 45

7	5	6	4	8		9	3	2
4	8		3			6		
2	9	3		5		1		8
	3		1				9	5
		9	5	3	7	2	8	4
			2	9		3		
1	2				3		6	
9	4	5	6	1			7	3
3	6	8			5		2	1

EASY - 46

6		9	5		7	8		2
		7		3	2	6	4	1
3			6	8	4			9
			1				2	5
2		8		7			6	3
	9	6		5	3			7
	1	5	8			3	9	
9	6	2	3		1	5	7	8
8	3		7	9		2		

EASY - 47

	6		2	3	8	5	1	
		5		7	1	3	6	9
3	1		9	5		8		2
7			8			1	5	4
1	9		6		5	7	2	3
	3			1	2		8	6
9		1	3	2	4			8
6	4	3						
		2	1	6		4		

EASY - 48

5	2				7			
1	7	3	6			5	8	9
	8	9			3	7	2	6
9	6	4	3		8	2	7	1
	5	7		9	1		6	4
2	3				6		9	5
		5	8			6		
	1			6	2		5	
		2	1	7	5	4	3	

EASY - 49

6				2		8	3	
4		8	1	6		2		
7	5	2	4	8	3		6	
		6		3	4	1	5	
3	4	5	6		8			
8		1	9	5	2		4	
9	8	4	3	7			1	
1				9	5			7
	2	7	8	4			9	6

EASY - 50

	9				4	5	1	
3	4		6	9	5		2	7
2			7		8			3
	7		5	6	9		8	
8	2		4	7	1	6	3	
1		6	3	8	2			9
9	8	7	1		3		6	
	1		9				5	8
5	3		8	2		7	9	1

EASY - 51

		6	8	2	3		4	
	2	4	6					
		3	4	7	9	6	2	1
1	3	2	7	4			8	9
6	7	8	5	9		1	3	
	4				8		6	
2		7	9		1		5	
3		1	2		4	8		6
		5		8	7	9	1	

EASY - 52

	9	4	1	6	5	3	2	
1	5		7				6	
2			3		4	5	7	
	2			4	3	6		7
8	3				6	9	5	4
			9	7	8			3
	6		8		1	7	4	2
	1		4	5	7	8	9	
	7	8	6				3	5

EASY - 53

8				7	6	3	2	
7		6		3	2	4	9	1
		9	1	5	4		7	8
1		7	2			9		4
5	9		7		8		1	6
6		4		9	1			
	7	8	3		5			9
9			4	1				
3	1	5		8		7	4	2

EASY - 54

2			3	1		6	8	
3	8				4	7	1	5
1		5	9	7	8	2	3	4
6	1	2		4	7	3	9	8
			8			4	7	2
7	4		2			1	5	6
9	3				2		4	
4		7	1	5				
8	5	1	4		6			

EASY - 55

	2	5	8	1				
	7		2			5	1	8
	4	1			9	7	2	6
1	9		6	4		8	3	2
4	3	8	7		2	6	5	1
	6			8			7	4
7	8		1	5			6	9
	5	3				1		7
			9	7	8	3		

EASY - 56

1		5		8	6	4	2	
3		9	7		2	1		
	2	4		5	1			
9	8	1		2	4	3	6	
6	3	2				5		
4				3		2	8	1
2		3	4			8	1	
7	4	6				9	3	2
		8	2	9	3	6	7	4

EASY - 57

	2		8		7	9	4	5
		4	5	6		7	8	3
		5	3	9	4	6		2
	7	8	1				9	6
	9		2			1	3	7
	3	1	6					
1	5					3	7	8
3			7	8	1		6	9
	6	7		5	3	4		1

EASY - 58

	1	4	5	2		3		9
			9	1	8	6		4
6	8	9		4		5		
9	4	8	1	5	3	2		
	3			6	9		1	5
1	6		4		2		9	3
5		1		8				2
4	7		2				5	8
		3	7	9	5	1	4	

EASY - 59

4	3	8	9	5			2	7
5					1	9	8	3
9	1	2	3	7	8	5	4	6
8				3	7	6	5	9
6	9	7	1		5			
				8				
7		4	8	1				
1			7	9	3	8	6	4
				6	4	7		2

EASY - 60

						4	1	
7	9	4		3	1	5	6	8
6	1	8	5		4		2	3
		3	6	2	5	1	9	4
5	2	1	7					
	4	6	8	1	3	2	7	5
4	8			9			5	2
			3	8	2	7	4	
			4		7			

EASY - 61

5	7			2				1
	6	8	9	4	1	7		2
1		4	7	5		3	9	
7	5				9	2	1	6
		6	5	3	2	8	7	
8	9	2		1	7	5	3	4
		5	2	6	8			7
	4					6	8	3
6		1						

EASY - 62

2	5		1		6		4	
8	7	6	3		9			
4	9	1			8		3	6
9	2	4	6	8	5	1	7	3
1	6	7				5	8	
	3			1			2	
			8		1	3		
3		9	4	5		8		7
	8	2			7	4	1	5

EASY - 63

			5	6		4		8
	4	3		2		9	5	6
7		6	4				3	
		8	2			7	1	9
	9		7	8	1	6		
3	1	7		4		2		
	3	4	8	7		5	2	
2	8			5			6	7
6	7	5	3	1		8	9	4

EASY - 64

6		7	9	4	3	8		2
3		5	2	7		6	9	1
	2	8	1					7
	6	9		1	5			8
2		3	6				4	5
5		1		2	4		6	3
7	3			9	1	5		6
1	9		5	6		3	8	4
		6		3			1	9

EASY - 65

3			5	2			9	1
4	1	6	3		8	2	5	7
		9				3	6	8
6	9	5	8	1	2	7		3
7	3		6		4			9
8	4	2		3		5	1	6
	8				5	6		
	6	3	4	8		1		
		7			3		8	

EASY - 66

		3	5	9	1	4	2	
5	4	7		8	2	3		9
9	1	2				6	5	8
1		5		6	9	2	3	4
8	2	9	1	4		5	7	6
	3			2	5	8	9	
7	5		2					3
		1	9			7		
	9	4			8	1		

EASY - 67

7			1	5	3	6	2	4
		4				5		
5	3	1	6	4	2			7
8	2		4				1	3
6	7	5	2	3	1		8	9
1	4		8	7		2	6	
9	8	6	5				4	
			9	8	4		7	6
4			3			9		

EASY - 68

	6			9				
5	1	8			7	4	2	
		9	5	2	1	7	6	
1	4			8	3		9	6
2	7	3			9			4
		6	2		5	3	7	1
3	5	4	9	7	6	8	1	2
6				1	2		3	
9	2	1						7

EASY - 69

9	4	6	1		3			
2	3		9		5	4	6	7
			4	2	6	1		9
7		8		9	1	3	4	
4	5	9	7	3			8	
		3		4	8	7	9	2
8	9		3		4	2	7	
			2				5	
3	6		8				1	4

EASY - 70

4		6	3	1	8	9	5	2
8	3	5	4	2	9		7	
9	2	1				8		3
1						7	3	
	8	9	6					4
3	6	7	5	4			8	9
	1	3	2	6		4		8
5	4		8	9	3	1	6	
			1			3		5

EASY - 71

	4	6	2	3		8	1	5
	8	3	9		1	2	4	6
	2	1		8		3		
3		7					8	4
	5	8	7		3	9		
	1	9	5	4	8			
	3		6	7			5	2
	9	5	8		4	7		
	7	2		1	5		9	8

EASY - 72

				2				5
			7	1		3	2	9
	7			5	9	1	6	8
	2	5	9			4		6
8	9	4		7				1
1	6	7	3	4	5		9	2
6	5	9	1	3	7	2		4
7	8			9		6	5	
	3	2	5		8			

EASY - 73

			9	6			1	3
6				2		5	7	9
9	2	1		7	3	8		6
		6	2		8		9	7
	8	2	4	9	7		5	1
7	1	9			6			2
2		8		4			3	5
1			6	8	5		2	
4	5	7		3	2		6	8

EASY - 74

		2		9	4	1	3	
9	6	3	7		1			2
4	7	1	8	3	2		9	6
7	1			8	6		2	
8		6		2			1	4
	3		4	1			6	9
3				6		2	7	
1		8	2	7		6		
	2	7	1		3		5	

EASY - 75

3	1	6		7				8
		8	6				1	
5	9	2	1		8	6		4
	8		7			3	5	
	7	4	3		5	8	2	
1			8	2	9			7
9	6	7	4	5		1	8	
	2	5	9	8		7		6
8		1	2	6		9		

EASY - 76

								8
4			7			5		3
8	1	2	3	5		4	9	7
7	8	4		9			3	
		3	5	7	4	9	8	1
	9		8	3		7	4	
2		8	9	6	5	3	1	4
1	3	6		4				
	4			8	3	6	7	2

EASY - 77

7	2		1	5	8		6	
6		5	9		4			1
1		8	3	2	6			7
4				6			7	
		1		3	7	4		
	7			8	9		1	
5				1	2	7	4	9
2	1	4	7	9	5	6		
9		7	8	4	3	1	2	

EASY - 78

3	5		9			4		1
	2	6	5		4	3		9
1	9	4					8	2
			1	2	3	9		4
9	4			7	5	1	3	
5	1	3	8	4		6		7
	6	9			8	2	1	3
			3		1	7	4	6
			2				9	5

EASY - 79

1	3	5			6	7		
9		4			8			1
	8			4	3		9	2
3	6	8		9	1		2	7
5	7							3
4			2	3	7		6	5
6	9		4		5	2	7	
8		1	6	7		3	5	9
			3	8	9	1		6

EASY - 80

				1	8	6		5
4				9	7		1	
	1		5	2	3	4		9
5	7		2	6		3	4	8
9	2		8		1			
	6	8				1	9	2
8		2	1					6
6	4	7	9	5	2		3	
1	5	9	3	8	6	7		

EASY - 81

		5	8		2		1	6
8			4		6	2		3
	6			7	1	8	5	
	7	6	5		4	1	3	
1	8	4	2	6	3	5	9	7
	5			1	7	6		8
	2	7		4	9	3		5
3			7	2		9		
	9			3	8			4

EASY - 82

		3	7	8	9	6	5	2
5	8	9		3		4		1
				4	5	9		
	2		5		1		9	6
7	1			9		5		
3	9	5		6		8	1	7
2	5		3	1	4	7		
		1	9	5	7	2	8	3
9	3	7	6	2		1		

EASY - 83

4	5				7	1	3	
8	3	7	6		1	9	5	4
			4	5	3		6	
1				4			7	9
	2	5		9		6		
9			3	7		2		
	9	1		6	5		8	
5	4		7	1		3	2	6
6	7	2			4	5	9	1

EASY - 84

9	8	7			3	4		5
4	3							2
2			4	7	8	9		1
	4	2	8		7		9	3
3	7		5	4		2	1	6
1		9	2	3	6	8		
5	1	6	9		2	3		4
8	2		7					9
	9	4		5	1	6		

EASY - 85

5	7	6		2		4	9	
8	4		5		1	2		
1	2	9				3	8	5
	9	5		4	8			3
6	8	4	3	1	9	5	2	
3		7	2	5		8	4	
	5	1		8	2		3	
	6	2		3		9		
4		8				1	6	

EASY - 86

	7		5	4			1	2
3			1	6		9	7	
1		4		8		3		6
	9		3				8	
		7		2	5	1	9	3
	3	2	9		8	4		7
9	4	5	6	3	1		2	8
7	8			9	4		3	
2		3		5	7			9

EASY - 87

9	1	8		2	5			6
		4	1	9	6	5		2
2				3	7		9	
8	4	1				2	6	9
	9			4	8	7	1	
6	5	7	2		9		3	4
		3	9	6		1		7
4	7						2	8
1		9	7		4		5	

EASY - 88

		1				5	2	
6	3	5		8	2	4		7
2	7			5		8		
	1	6		4	9	7		8
9	8	7			5	2	6	4
4	5		8	6	7	3		
	2	9	7			6		3
7		3		9			4	2
1	4	8		2	6			5

EASY - 89

2			4	5	3		6	
6		5				4	1	2
	9		2			3	5	8
9			1	7				3
		6		3		7		9
3	5	7	8				2	
8	4	3	7	2	5	1	9	6
	6	2	3		4	8	7	5
5	7	9					3	

EASY - 90

3	2	9	6	5				8
		7						2
8	6		7	3	2			
2		8	4	6	9	5	3	1
	3		2	1	8	7	4	
6		4	5		3	2	8	9
	9							3
1	4	3		2	6	9	5	7
			3		7	6	1	

EASY - 91

9	1		4		5	3	8	7
	2	7	8		6		1	4
5		8	7	1	3		6	2
		9	1			8	3	
	6	5	9	3	2	7		1
1	3	4		7			9	5
4			5		1	6	7	3
		1		6	9			
6	8					1	5	

EASY - 92

2	9		3	6				
1	3	4	2			7	6	
5	6		4	7	1			3
4	7						9	2
	2				5	3	8	
8	1			2	6	5	7	4
6	5		1		2			
	8		5	9	4	2	3	6
3	4			8	7		5	9

EASY - 93

7	4	8			9	3	5	6
5	1		7			2		9
3			4	6	5	7	1	8
	5	7	9	3		6		1
6		9	5	7	1	4	2	
1	2	3		4	6			5
9				2		8		4
	7		3	9	4	5		
	3							

EASY - 94

1				5	2	7	9	6
5				9	6	2	1	
2	9	6		8		4	3	
6	7	2	5		1			9
9	4	5	6	7			2	
8	1		9	2	4	6		7
	5			6			7	
	2	1	8	4		9	6	
4				1	3			

EASY - 95

2	3	7		4	9		8	
	6	5		8	1		3	7
	9		6			5	4	2
3			4	6			5	
6		9				8	2	4
	8		9	2		3	6	1
	7			5		4	9	3
		3	1		4	6		8
	4	8	3	7		2	1	

EASY - 96

6	4		3	2				
	7	1	6	4	5			8
	3		1	7	9	2	6	
		8	7	6		3		2
	1				2	6		9
4	6		9	8	3	7	5	
	5	7	2		4		9	6
1	8	6	5			4		
9	2		8				7	5

EASY - 97

	7	6	1		3	9	8	5
1	2		7	8		4		6
	3	9			5	1	7	
7		4		6	8	3		1
9		8	3		2	5		7
		2	5		4		6	9
		7		5		6		
	9	1	8	3	7			4
	4	3	2				1	

EASY - 98

3			2	8	9	4	5	
2	9		4		1	3	6	7
	4		3	6		2	9	8
	3		1		6	9		5
		7	9		3	1	2	6
	6		8				3	
5	8				4		1	
	2		6	1	8		4	9
6		4	5				7	3

EASY - 99

4		2	5		8		9	1
	9		3				8	4
8	1	3			6	5	7	
		9			3		2	
2		7	8		9			
6		1	2		7	4		9
		4	6	3	5	7	1	
3	5	8	7	2			4	6
1	7	6	9		4		5	

EASY - 100

4	8		3	9			2	1
1							5	9
			2	1		8		4
9	1			2		5		6
	7	8	6		1		4	3
		3		4	8	2	1	
7	3		4	8	9			5
	4	1	5		2		9	8
8	5	9	1	6	7		3	

EASY - 101

	2	5		4	8		3	
3	9		1	2	7	8		
8	4		5	9		1	2	
				6	5	4		
4	5		7	3	1			2
	7		9		4	5	1	
			4	1	9		5	8
5	8	4	3	7		6	9	
2	1	9	8		6		7	

EASY - 102

	3		7		8	5	4	1
			9	3		2	6	7
		7	5	4	6	3		8
			6			1	8	
	6			1	3	7	2	
	4	1	2		9		5	
7	9		1			4	3	6
2			4	6	7			5
	1	4		9	5	8	7	2

EASY - 103

	7		5			8		2
5	9	8	1					7
3	4	2		6	7	9	1	5
6			4		8	5	7	9
7	5		9			3	4	8
		4	3		5		2	
				8	4	1		
4			2	5	3		8	6
8	3				1	2	5	4

EASY - 104

1	5	6		3			4	
8	7	2	5	4		3	6	9
3			8				5	1
9			4	1	3		8	
5						2	3	
6	8			5	7	1	9	
2	3	8	1				7	5
4			3		5			8
	9	5	6	2	8	4	1	

EASY - 105

6	3	7	4		2		9	1
	8		5			2	7	
		2	6	9	7	3		
7		3		4	8			5
9		8		5		4	1	
		1	7		9	6		8
8		6		3			5	9
2	9	5	8	6		7	4	3
		4	9		5		6	

EASY - 106

			1	2	8	4		9
		1				2	6	
2	8				6	7	1	
8	1	3	7				9	2
7	9	4	2		3	1		
		6	8	1	9	3	4	7
	4	5	3			6		
	6	8	5	4	7	9	2	3
3			6				5	4

EASY - 107

3	2	5			7	1	6	9
	7	4	3	6	1	8	5	
6	1	8	5	9				4
2		3			8		1	5
	8		2	5	9			
7	5		1	4	3	6		8
	9		8	1	6	5	7	
5		7	9		4	2	8	1
	3				5	4		6

EASY - 108

2	5	7	9		4			6
4	6	3	5	8		1		9
		8	3	7	6	2	4	
	1	2	4		5			8
	3	4	8	2	1	5	9	7
5	8		6				2	1
			7	6	3	8	1	
8							6	
		6			8	9	5	

EASY - 109

8		4	6		3		2	1
7	2	1	9	5	4			
	6	5		8	2	7	9	4
4				2	9	6	7	5
	7	2	5	4	8		3	
9	5		7	1	6	4	8	2
5	4	9			7			
	3				1	8	5	
1				3				7

EASY - 110

			9	8		3	2	4
9	8	2		6	4		5	
5		4	1		2		9	8
	6	9	8	5			7	1
3	2	1		9	7			5
7		8					3	9
8			5	2	9		1	
2	7	5		4	1		8	3
1				3			4	2

EASY - 111

2	1	6			8		5	
3				4	5	8		2
	4		2		1	7	3	
	2		8	1	3	6	7	
	8	3	7	2			9	1
	6	1	9	5	4	3	2	
9	5		1			2	8	
6	7	8	5		2		4	
	3			8		5		

EASY - 112

3	1			9		7		
4			1		7	2	3	8
2	7						9	1
			5		9	1	8	3
9		7			1	5		4
1	3	5	4		6	9	7	2
	4	2	9	1	3	8	5	
	9	1		4	5	3		7
					8	4		9

EASY - 113

9	3	1	6	4	2	5		8
2	4	8		5	7	3	1	6
5	6	7	8	3	1			2
		5	1				4	
4	9		7	6	5		8	
1	7		4	8		9		
6		4			8		3	
7	2	3				8		9
	1				6			

EASY - 114

9		2		5	8			6
7			1	6	9	8	3	
8	6			2		1	5	9
	2		8	9			4	
5	8	6	2	4	3	7	9	
3		4	5			6	2	
	3	9		7	1			
6	5		4	3		9	1	
2		1				3		4

EASY - 115

4	5	6				9	7	8
8	3		6		9			2
	9		5		8	1	6	3
	7			6	1		3	5
9	6		3	5				1
3		5	2			6	9	
6	8		7	1			5	
5	2		4		6	8		
1	4	3		9	5	7		

EASY - 116

1	7	8	2	4		3	5	9
5	9				3	4		2
	2	3	8	9		1		
7		1			4			
2			5	7	8		4	
8	6		9			5	3	7
6		5		8		7	9	3
9	1	2	4	3	7	8		
3			6					4

EASY - 117

7	8				4		5	3
2		9					6	7
		6	5	9	7		8	1
9				6		7	3	
5		3	7			6	2	8
	7	8	4	3	2		9	5
3	2	7	9	5	1			6
8		4			6		1	2
1	6	5					7	

EASY - 118

3	8	5			6	1		
1		6	3	9	7		4	8
7		4	5	8	1		6	
9	5		6	4	2			3
4	7	3	1		8			6
2			7	3	9		8	
6	1			7	3		5	
5							2	7
			2	6	5	9	3	

EASY - 119

8		7		9	5			6
	9			7	8			5
	1		4		2		8	9
4		5		8			3	2
6	8		5	2	3	4	7	1
	3			4		5	9	8
9	6	2	3					4
	5	3			4	9		7
7	4	8	2			1		3

EASY - 120

4	5			9			6	
3		8						
9	2		4	1	8			5
	8	3	9	2	4	7	5	
7		2		6		9	3	1
6	9					8	2	4
	6	9		8	2	1		
8	3	4	6	7	1	5	9	2
			5	4		6		3

EASY - 121

5	7	1		9				
	9	3			4	7	1	
4	6		3		1		2	9
9	3		6		2	8		
	4		8	5	3	1		2
8	1			4	9	6	3	5
1	8		9	3	5			
	2	7	4				5	1
6			1	2		3		4

EASY - 122

	4	3	2	1	6			5
6	9	5	3	4	8	1	2	7
	1			9	7		6	3
		6	7	8		2		
4					1			
1	2			3	5	6	7	9
9		7				8	5	4
2		4	9	5			1	
5	6	1	8		4			

EASY - 123

	9		1	3			7	6
3	2		7	8	9	4	5	
			6		4	2		3
	6		8	1	7	9		
7		1	3	9				8
9			2	4	6	7	1	5
6	3	9	4	2	1			7
		8	5		3	1	4	9
	5	4				3		2

EASY - 124

	5	8	9		1		4	
	4	6	8		3		1	9
		9			4	6		
	8	2	4	7	6		9	
9			3	1		8	2	6
6	1	5	2		9	4	7	
4	2	7	1		8		6	5
			5			7	3	4
5		3				1		2

EASY - 125

6		8				7	5	2
	2		1	7		6	9	4
4			2	6	5		8	1
9		2	7		6	1	4	5
	4		5			8	2	
	8		4		2		7	6
	6	9			4			7
3	5	1	8				6	9
2			6	1	9		3	

EASY - 126

		2	7		9	5		1
4		1			6	3		
9		7		4	1			6
	9			6	5		3	7
5	2		3		7		6	8
3		6		1			5	4
2		9	1		4	7	8	3
7			6	2	8	4	9	
	4	5	9	7				2

EASY - 127

						9	4	3
1	3	5	9		4	7	2	
			2		6			1
6	7	3			2	8	1	
	5	1	3	6	8	4		7
		9	7	1		6		
3	1		6		9		5	4
7		2	5		3	1	6	8
5		4	8		1	3		9

EASY - 128

6	8	7		2	9	3		4
				6		8	7	1
	1	5	8		3	6		2
9	7			3	5		4	8
		8		4	1	7	2	6
	6			8	2	5		
7	9	1		5		4		3
		6		9		2	1	
8	2			1	7			5

EASY - 129

9		4		8		5	7	
3	7	6	5			1		
			2	7	6	9		4
		9	6	4				8
6			9	1	5	2		7
	5		7	3		6	9	1
8		3	4	2	9	7		
2	4	1					6	9
5		7	8	6	1		2	

EASY - 130

	2			9	3	7		5
9	1		6		7	2		8
6	7	5	4	2	8			9
3	9				2		1	7
		1	9	7	4	3	5	
	5	7	8	3		6		4
				1	6	4		3
	3	6	2	4				
	4		3	8	9			6

EASY - 131

	4		2	5	9		7	1
8				1		9	6	
9		5	8		7	4	3	2
	5	8	1	7	4	3	2	9
		4		3	2		5	8
7					5			6
5	3			2			1	4
4	6	7		9	1		8	
2		1	3					7

EASY - 132

4	2	8	9				3	5
9	1			5	2		7	
7		6		3			1	9
	4		2	8			5	7
2	7		4		5	3	8	6
8		5	1		3	4	9	2
	3		5	2		9		1
	8				9	5	6	
5		4			1		2	

EASY - 133

			1	7	4	8	6	3
3	6	4						
			5					4
	1	5	7		3	6		
6		3		2	5	1	9	
	2		8		1	3	4	5
8		6	2	4	7		1	9
		1			8	4	7	6
5	4	7	6	1	9	2	3	8

EASY - 134

		7	1	5	6	4	8	3
4		8		7			5	
5						7	2	9
		9		3		1	4	
7	6	4	8	2	1	9	3	5
			4	6		8	7	2
3	4			9			1	
6		1	2				9	
8	9	5	7		3	2		4

EASY - 135

1	2	4	7	5	9		6	8
8	7					1	5	4
	3	6	8			7		9
	1	7	5		8			6
3	8	2		4	6		7	1
6		5		7	1	8	9	3
	6					4	1	
		3	1	6	2		8	
				8	7		3	

EASY - 136

	1	5				7	3	2
2			1	9	7			
6		4				8		
5	6	2			4	9	1	3
1						4	8	5
		8	3	1				6
	4	6	5		1	3		
3	2	1		6	9	5		7
7	5	9	4	3	2	1	6	8

EASY - 137

3			2		7	8		1
7	9	5	8	4		3		
	8		9		6			
		9	5		4		3	
	7			8	2	9	1	4
4					9	5	6	8
6	5	7	4	9	3		8	2
9	1		7	2			5	
2	3	8		6			7	9

EASY - 138

7	5	3	8	9			4	6
	9	2			4	7	8	
8		6			7		2	3
3		9			8	4		
4		8	6	7	5		3	9
		5		3		6	1	
	8	4	5	2				1
	3	7	9				6	
	2	1	7	8	3	5	9	4

EASY - 139

3			9		1	2	4	6
7		6	2	3	4			1
1	4	2	6	5	8	7	3	9
8	1			9	6			
5						9	1	
2		9						5
		7	4	8		1	9	3
9	3	1	7		2			
	5	8	3	1		6	2	

EASY - 140

6	3	8	9	7	1		4	
7		4	6	8	5	1		9
5			3	4			7	8
					7	8		
	6	7	1				2	
9	4	3	8			5		7
1	7	2			8		9	6
4		6	7	1	9		5	
3		5		6		7	8	

EASY - 141

7	8			4	9	3	6	2
	6		1		2	8		
2	4	3	6				9	1
4		5		1		9	2	8
3	1	6		2				
8		2				1		6
6	3	7		9		2		4
1	5			7		6	8	
	2		4	6	5	7		3

EASY - 142

1	9		2			7	4	3
2	7	5	3	9	4	6		1
	4		1			2	9	5
	2	1	4	3				
7				8	2	4	1	
5			7	1	9	3		2
8					7	1	2	
		2		4		8	3	7
4	1	7	8	2	3			

EASY - 143

6	9	8	3	2		5		
		3	8		5	2	9	6
4	2		6	7	9		3	8
		4					6	2
2	5		9	8		4	7	3
8				3	2	9	1	5
5	8	6	7	1	4			
7	1							4
	4	9	2		8	7		

EASY - 144

	9			7	1	2	6	
6		7	9	2	4	8		1
1	5	2		8	3	7	9	4
	4	9	2	5		1		
5		1		3			2	9
7	2					4		
9	1	3	7	6		5		
4	6	8		9	5			
2		5				9	1	

EASY - 145

3		8	5		7	2	4	1
9	5		6	2	1		7	
		2	3	8	4			
5	7	6	9		2			
		9	1			7	3	
	4		7	6	8	9	2	5
4						3	9	7
8	3		4	1	9	5		2
6			2				1	8

EASY - 146

4	7	2	9	1	5		3	6
9		3		2				5
5			7		3	9		
	5				2		8	
7	2	9		8		6	4	3
8	3		6			2	5	1
2			4	3	6		1	
		6	2	5		3		
3	8	5	1		7	4	6	

EASY - 147

1	3		8		6	5		
	8		3	7	5		9	1
7	5			1			3	
5	2	1	6	9	3			
6		8	1				5	
4	9		7			2		
3	1			6	7	9	8	
8	6	7			4	1	2	
9	4	5		8	1		6	3

EASY - 148

2			4		3			
7			9	5		2		4
8	4	9		2	6		5	
	5	7	2	1	9		3	8
6			3	7	4	1	9	5
9	3		5		8	4	7	
				3	2		1	
	9		1	4	5		2	7
1	8		6		7	5		

EASY - 149

	9	3	8	7	1	2		
	1		5		2	9	3	
6	2				9	7	1	
4		9	1	2	5	6		3
3	5	6	7	9	8	1	2	4
	7	2	4		6	8		
8							9	7
	4							
2	3		9	8	7	4		1

EASY - 150

		2	5			6	4	
			4	9	1	8		3
3	8		7	6	2			5
	7	6	1	5	9	2	3	
2	4	1	6	3		5		9
		5		4				1
4	5	3		7			1	2
		8	9			3		6
6	1		3		5	4		8

EASY - 151

	2				6			4
5	1			2			9	7
9			5		3			1
1		8	7		2		6	
2	9		6		1	7	4	3
6	4	7	3	5	9			8
	6				8	4	7	
8	7		4	6			3	2
4	5		1		7	9	8	6

EASY - 152

	6	5		8	2			
	1	7	6			8	4	2
8		2	4		3	6	5	1
	5	6				4		
7	3	8		4	1	2	6	5
1		9			5	7		8
	2	4		1	6	5		
	8		5	9			2	4
	7	1		2	4	3		6

EASY - 153

		3	7	2		9	4	
				1	9			
5				6	8	2	3	7
3		6		4	2	7	9	
2	9	1		3		5		
4	7	8	9		6			2
6	4	2	5	7		1		9
9	3	7	6	8	1	4	2	5
	8	5		9			7	

EASY - 154

7	8		4	1		3	9	
2	9	4	3	5	8		7	
		1			7		5	8
1	5			9	2	6	8	4
9	2	8		6		7	3	
			8	3	1	9	2	
5	4		1	8	3			
8	7				9	5		
			6	7	5			9

EASY - 155

8				4		1		5
		1	5	8	9			3
	4		1		7	8	2	9
7	1			9	5			
	2	3		1		5	8	7
5					3	9		
1		2	7	3	8	6	9	4
6	7	9	4	2	1			
		8		5	6	2	7	1

EASY - 156

5		9		3	4	1	2	6
	2		8				3	
3		1		9		8	4	7
1	7	5	4	8		6	9	
			5	6		2		4
2	4	6		7				
		3	1		7			
	5			4	3	7	1	9
4		7	9	5	8	3	6	2

EASY - 157

6	3		4	5			8	1
4	8	5	6		1	7	2	3
9		1	8		7	5		6
						1	9	5
2	1	9	5		8	4		
			9		4			2
3	7	2				6	5	
5		8	3	7	6	2		9
		6	2	8		3		

EASY - 158

	6		7		8	3		
7	9			5	6		4	2
			9		1			7
3	5		2		7		8	
8	7	6		3			9	
1	2	4	8	6	9			
6	8	3			5		2	
2	4	5	6		3	1		8
9	1		4	8		5	6	3

EASY - 159

4		8		6				2
2		6			9	7		
			7	2	8		6	4
7	2	4	6			9	8	3
6	8		2					5
	1			9	4	2	7	6
1	6		5	4	7	8	3	9
3	4		9		2	6		1
8			3	1	6			

EASY - 160

1	4		6			7	8	
5		8			4	9	6	3
		3	8		5		2	1
8	2				7	1	3	
7		5	9			8		2
		6		8	2		7	
6	8	4		9	1		5	
3	5		7	2	8	6		
2		7	5	4	6	3		

EASY - 161

	5	4	2	7		8	6	3
6	8	9		4			5	
	2	3			5	9		
9		8	4			3	7	
4	3	7	1					
2	6	5		3	7	1		
		2	5	9		4	8	1
8		1			3	5		6
	9		8	1	4	2	3	

EASY - 162

1		8	2	3				9
5			1		8	7		
			4				1	8
	1	7	5		6	9	2	
3	5	4	9		2			7
2		6	8	7	3		5	
		3	6	2	5	1		
6	7	1	3			2		
4		5	7	8	1	3	9	6

EASY - 163

3	5	9	6	4	8	2	7	1
7	6			2				
	4				5	6	9	8
	2						6	
	7	4	2		9	8	1	3
1	8	3	4	7				2
	3	2		9		5		
		7		6	2	1		4
	1	6	3	8	4		2	9

EASY - 164

					3		6	7
5	7			4	2	9	8	
6	1	9	7		5	3		2
7	5		1					9
		6	3	5	4	7	1	8
8	3	1	2			4		
2	9	8	5	3	1	6	7	4
1	6	5		2				3
3			8	9	6	1		

EASY - 165

	6		8		3	9	2	1
2	5		7	9				4
		9						7
4	8	5	6	1	9			2
		1	3	2				
3	2			4	7	1	9	
			1	7	6	2	4	
1	9		2	3	4	8	5	6
6		2	9		5	7		3

EASY - 166

	7	3	4	2	6			5
	5	2				4	9	
6	8	4	1			3	7	
3	2		9	8			1	
	4	9			1	5		8
	6	1		3	4	9	2	7
		7	6		2			3
	9	8	3		5		6	1
	3		7	1		2	5	

EASY - 167

8			4	9		5		3
9	1	6	3		8	4	7	2
		4	2		7		8	9
1		2				7		8
	4	9		8	2	6		1
	7	8		3	9		5	
7					3	9	4	
	6	3		1	5			7
2		5	8	7	4	3		

EASY - 168

8	5	6	9	7	2	1		3
2			4	1			5	8
4	3		5		6	7	9	2
1					9		7	
	8	3	1		4			9
	2		7	6				4
		2	8	9		4		1
3				4	7	9	8	6
9			6			5	2	7

EASY - 169

9	5				1	3	7	6
1		3	6	9	7		8	
7	6	2			5			
	7							
2		4		6	3	8	9	
		6	5	1	2	7	4	3
5	8			7	6		2	9
4		9				6	3	7
	3	7	9		4	1	5	8

EASY - 170

		2	1	4	6	3		
3	4			2				5
1	6	8	5			4	9	2
	2	4	7	5		8		1
8		9		6		5		4
	1	5	4	8			2	6
		3	6			2		8
4	8		3			7		
	9	1	8	7		6	4	3

EASY - 171

3	7	1		5	8	2	9	4
5				3	9	6	7	
		9	7	1			5	
	5	4	3	7	6	1	2	
1		6			4	7		
7		8	2	9	1	4		
2	4	3				9	8	
6	9					5	1	2
		5	9	2		3	4	6

EASY - 172

	2		8		3	1		
8			4	5			2	9
	1	3	7	9		6	5	8
3		1		4	8	9		7
	8	7		1	6	5	4	3
6				7		8	1	
	9	8		2	4	7	3	
7				3	9			
1	3		6		7	2		5

EASY - 173

6	2	7	9	5		4	3	
1	3			4			6	2
	8		6				5	7
		9	4	1		3		5
4	5		3	8			7	9
			5				4	
		6	1		5	8	2	
5	1			6	4	7		3
	4	3	2	9	7	5	1	6

EASY - 174

9	2	1	4		3	5		8
		3		5	8	2		4
8			1		2			6
		5	6		7		4	2
	3	9		2			5	7
	7	2	3	1	5			9
	1	8	2				6	
	4	7	5		6	8	9	1
5	9		7	8		4		

EASY - 175

2	5				7	1	9	6
			4	5	6			8
	3	6		2	9	4	7	5
	1		6		2	9	5	
	2	3	7	1			8	4
		5	3		8	2	1	7
	6		5			8	4	2
5	8		9	6			3	
	7				3	5		9

EASY - 176

	8	5	1	9	7			
2		3	5	4	8	6	7	1
	7	4		2	3		9	5
		8		5	6	3		
7		1		3			8	2
		2		7		4	6	
4		9	7		5		3	8
8	3			1				
5			3	8	4	9	2	6

EASY - 177

		4	9	3				
7	9	2						3
	6	3		8				1
	5			9	8	7		6
4			2	5	7		1	
	7		4	1	6	3	5	2
8	2	5	6	7	1	9		
3	4	7	5		9		6	
	1	9		4	3	5	2	7

EASY - 178

	4					1	6	5
6			7		5		2	3
3	1		4		2	8		9
5			8	9	1		4	
9			3	7	4	2	5	1
4		1		5	6			
		6		4		5	9	
7	5		6		9	3	1	4
1	9	4	5	2				7

EASY - 179

	9		6			5		
8		7	9	1	5	6		
4	5	6		8		2	1	9
	3		4		8		9	5
9	8	5	3	2	1			
6	1	4		9			8	2
1	4	9	8		6	7	2	3
2		8				9	5	
	6	3	2	7	9		4	

EASY - 180

	7	3		8			1	
2	1	8		4		9	6	
4		6	2	7		5		
3	6	9	1	2			7	
	5			9		1		
1	8	4			7	2	3	
6		1	9	3		7	5	8
8	2			1	6		9	4
	3			5		6	2	1

EASY - 181

9	1	7	6		5			2
8	3	2	4	7			6	9
6	5	4	8				1	7
2	8	5	7	6	3	4		
4	7	9	1					3
1	6	3	2		9	7	5	8
							7	6
			3	1	7	2		4
					6	1		

EASY - 182

7		8	3					2
	2		4	8		3		5
1	3	5		9	2			7
3	5						8	6
6		2	8	5	1			3
4	8	7	2	3	6	1	5	9
2	6		1	7		5	3	4
		4	5	2			9	
5	1	3						

EASY - 183

8	7				2	6	9	
	3	6	8	4			5	2
1	2	5	3		9			
7	4	2		9	6			1
6		3	7	8	1		2	5
	1				4		6	
		9	4	7	3		1	
				2	5		7	
4	5	7	9	1	8		3	6

EASY - 184

9	1	4	7	6	3	8	2	5
		5	4	9		6	3	
6	3						4	1
		8						
5	9	6		3	8	1	7	
		3	9	7	5	2		6
		1		2		7	5	
3	8			1		4	6	2
	5	2		8	4	3	1	

EASY - 185

3	1	7	6		8	4	5	
8				1	5			
5	2	9		4			1	6
	6	5		3		7		2
7		2	5		1	9		
9		4	8	7		1		
	7	1		5	4			8
2	5	8		9		6	3	4
4			2		6		7	1

EASY - 186

4		8		3	1	7	2	
2	7		8	4		5		1
9	6	1	2					3
7	3	4			2	1		9
6	9	2	1	7		3	5	8
8		5			3		7	
5				1	9	6		2
			7	2	8		3	
3		9	4	6				7

EASY - 187

			1	3		5		8
7			5		9	1	2	3
	3			2		7	9	4
	1	4	3		8	6		2
8	6	7		1				9
5	2	3	6	9	4	8	7	1
3	5		4			9	1	7
		9	7	8			3	
		1			3		8	

EASY - 188

2	3	5				7		9
9				7		2	6	5
8		7	2		9	3	4	
	5		6				2	3
		8	4		3	6	5	
					2	8	9	
3	9	6	7		5	4	1	8
5	1	2	8	3	4	9	7	6
	8	4		6	1	5	3	2

EASY - 189

	7		6	2	1	5	8	3
6	1			5	3		9	4
2	3		8	9				
1		6	2			8	7	
7	8		5	4			3	
5	9		1		8	4		
3	5	1			7	9	2	
9	6	4		8	2	7	5	
8			9		5			

EASY - 190

	8	6	9	4		7	5	
5	3				7		6	8
2						1	9	
6			2					7
9	4	8	7	3			2	1
	2	7	8	6	1	9		
4	1		3		9	2	8	6
8				1	2	5		
	9	2		8	6	3	1	4

EASY - 191

2	5		9	7			6	3
	6			1		4	5	8
3	4		8			9		2
5	9		3	2		6	1	7
	7					3	8	5
	1	3	7	6		2	9	
1			4	3		5		6
			5		2	8	4	
	2	5	6	8		7	3	

EASY - 192

	9	2		5		4	8	
5		8		2		6		7
	6			7	8		2	9
3	7	9		1	5	8	4	
1	5	6	4	8	7			2
8				3	9		7	
2	8	7	5		3		6	
9		5		6	1	2		8
				9	2	7	5	4

EASY - 193

				7	2	3	9	
	7		1	5			4	8
4					9	7	5	
1	9	3			8	4	7	2
6	8		7					9
	4		9	1	3	8	6	5
	1	2		9		6		
8	5	6	2	4	7	9	1	
9	3	4	6			5		7

EASY - 194

1		5	7		9		3	2
		9	6	3	2		4	
	4	2		8	5			6
4	7	1	8		3		2	
8			5	9			1	
9	5	3	2	1	4		6	
5		4		7			8	1
6			9	2		4	5	7
	1	7		5	8	6		

EASY - 195

9	8		1	2	6	7	5	4
1		2	3	5	4	8	6	
	5	6	7			2	3	
6	4		2			5		
3		5			8	4		7
7	2	8	5			1		6
5	9				2			
2				9	1	3		5
	3	1		7			4	2

EASY - 196

	4		5	2	6	9	3	
2			1	4	3			
3	5		8	9	7		2	1
	7		4		1	5		
4		5	2	3			1	6
6	1			8			4	9
	3	9			2		5	
5	2		9	1	8	3	6	
8			3	5		1	9	

EASY - 197

		7	9	8	1	5	3	4
5	1	3	7	4		9		
4		8	3		2	6	7	1
2	4	1		9				7
8		9		3	4	2		
		5		7	8		1	9
7	8	4		2				3
1		2					9	5
9	5		8	1	3			

EASY - 198

	4		2	5		7		
7		2		9	8			5
5	6	3	1	7	4		2	9
3		5		4	2	1		6
		9		1	5	4	7	3
	7			3		5	8	
		7		6	1	9		8
8	9			2	7		3	
4	5			8		2		7

EASY - 199

2		4	3		8	9		5
9		6	5	4	2		3	
3			9	1	6	8	2	4
4		8		9				
7						2	4	1
	6		4		3	7	9	8
1	4	2	7				8	3
6	3			8	4	5		
8				3	1		7	2

EASY - 200

	4		7	9			3	6
6		5		4	3	9	1	8
9		3	6	1	5	7	2	
	5		9				8	7
	2	8		6	1		9	5
4		9	8				6	2
2		4			6		7	1
5				8	2		4	
8		1		7			5	3

EASY - 201

3	4	6		9				1
2	7	5		3				9
9	1	8	2	4	5	3	6	7
	8		5	7		4	2	6
	6	4	3		2			
				6				5
8			6	2	1	5	9	4
	5			8	4	7	1	3
	9	1	7			6		2

EASY - 202

	5			3		9	4	2
				7	4		6	
4	6		2	9	1	5	3	7
		1	4	5	3	8		
5	9		1			6	7	4
	2	4					1	
7	8	2	3	1				6
	1	6	7	4	8	2		3
3		5	6	2		7		

EASY - 203

4		9	1		8	6	3	
3	7	6		9	4	2		1
8	1	2		6	7		4	9
		4	8		2	1		3
	2	3	4	5	1		6	
	8		6	3			2	4
	4	8	9		5		7	2
		1		4		8	9	
5	9		2				1	

EASY - 204

3	2	4	9			1	7	
	7			8	1			
	8	6	2		3		5	9
	1	5					8	7
2			8	1	6		3	
8		3			9		1	
	5		1				2	6
9	6		5	3	8	7	4	1
4		1	6	2	7	8	9	5

EASY - 205

	3	5	2		8		7	9
2	4	9	3	6	7	8		
	7	6	9	1	5			3
5	8		7	3	9	1	6	
			8			9		4
9			4	5	1		3	
		8	1		3			
6		1			4	3	9	7
4	9				2	5	8	

EASY - 206

6		1				3	5	4
		2	4	8		6		
	3		1	5				
					8	4		7
2		8		7	4	9	1	5
5	4	7		9	1		6	3
	9	6	7	3	5	1	4	
1	2	5			9			
3	7	4	6	1	2	5	8	

EASY - 207

5	7	9		6				8
	8	3	7	9		2		6
2			5	8		9		3
	9			2		5	6	1
	1	2	4	7			3	9
			6			7	2	4
	6	8	9	3	2	1		
3		1	8	4		6	9	
	2		1	5	6		8	

EASY - 208

	4						1	
	1		3	4			6	9
	3	6	1	5		7		
3		4	8	7	1			6
5	6	1	9	2	4		8	
9		8			3	1		2
1		7	6	3	2	4	9	
	5	3		9	8	2		
4		2			5	6	3	8

EASY - 209

9	5			6	8	1	4	
3				5	4			
4	6		9	7				3
5		6	7			8	3	
	9		4	8	6		1	2
2	8		5				7	4
1	3			2	5	7	9	
	2		1	4	7	3	8	5
8	7		6		3			1

EASY - 210

4	1	8	3	6	9	5		
	7	6	4	1	2		3	
3	9		7		5		4	6
	5				6	3		9
2	4		8		1	6	7	5
6		3	5	9	7	4	1	2
8	2		9	5	3		6	
9		7		2				3
					8			

MEDIUM - 211

		6			9		3	4
7				4	3	2	9	1
9			2	1		7		
	2	9	4				8	5
6		5						7
4		1	5		8	3		
5		2	7		6	4		
	6		9		4	8	7	2
8			3	2	1		5	9

MEDIUM - 212

		8	6	7		9		4
2		6	4	3		5	8	1
4	3	9		1	8		7	2
			8		4	3	2	6
8	6		7		3	1		
	2		9	6		8	4	
			3	8		7		9
7			1		5			8
		1	2		7		5	

MEDIUM - 213

1		5	9	4	3		2	7
2	6		5	7		9	8	4
4	7	9	8					1
	2	8			5	3		6
3		6	2	8		1		9
9		4	3					8
	9			5	8	4		3
5	3		4		9		6	2
8		1		3				5

MEDIUM - 214

	4	8			2		5	
6			8		4		2	3
7	2	5	9		1	6	4	
1	5		3	9	7	4		2
		3	2				7	
9	7		4		6	5	3	
8	3	7	1		9		6	
2		4			3		1	7
	6		7				9	4

MEDIUM - 215

9		6	1		5			
		7	9	6		5		
1		2	4	3		7	6	
3		5	6	2	1			
		8	3		4		7	5
		9	5		7	2		3
			8		3	9	5	6
8	6	4	2			1	3	
	9		7		6	4	2	8

MEDIUM - 216

5						2	3	
2			3	7		6		1
1		3				7	9	5
	9	7			6	5	8	
3		8	2			9	7	
6		5		9		4		3
		2	9	6	1	3	5	7
	5	6	7	8	3			
		1	5		2	8	6	9

MEDIUM - 217

8		1				6		5
6	2	3			9		4	7
4	5			6		3		8
	9	4	7	1	2		6	3
	8				5	7		
	1	7	3	8	6	4		9
	6	2	4	9	8		3	1
			6			2	8	4
				2	3			

MEDIUM - 218

2	3	8	4	1	5		7	
4	5	7		6	3	1	2	
6				7	8			3
	7		6	3	9	2	8	4
9	2		5	8	4	6	1	7
8	4	6			1			
					7		4	
1		4						9
7	9	2	8				3	1

MEDIUM - 219

9	3	4	2	5		1		8
		1	4	8	7			
5	8		3	9				
	7		1				3	2
1		8	9			7		
	9		7			8	6	1
8	5	3	6		9	2	1	
			8		4		9	7
7	4	9			2			

MEDIUM - 220

		1	7	6		9		
	5		4	8	9	6		1
	9	7	2		1	4		3
1	6		9	2				5
8	7	5	1	3	6	2		
2	3							7
			3	9	2	7	1	8
9		8	6	7	5			4
7					4	5	9	

MEDIUM - 221

3			2	1		6	4	5
2			7	4				3
	5	1			6	2		8
			1	8				7
7			6			8		
5	8			7	3	1		6
9	7			3	1	5	8	2
8		3		2				1
1		5	8	6	7	3	9	

MEDIUM - 222

7	2	8			6	9		5
6					2	7	3	
	9	4		8		1	2	
		9		7			1	3
2	3		6		5	4		7
5	4	7	3	1	8			
	8						5	1
9	5	3	4		1	8	7	2
	7				3		9	

MEDIUM - 223

8	4	7	6			9		2
3	5	9			2	6	1	4
	1		4	9	5	8	3	7
		1	3			7		
				2	8			
5			7		9	4	8	1
4	3	5			6	2		8
1	7	8		4		5		
6	9		5		7			3

MEDIUM - 224

5	6		1	3	4			
							1	6
2		7	6		8	9	4	3
6				9	3		7	4
3	8	4	5	7	1			
	7					1		
		5	8	1	2	3		9
1		6		4	5	7	2	8
	2		7		9	4		1

MEDIUM - 225

		3	5		2	4		
6			1	3	8		2	5
	5	7	6	4	9	3	1	8
7			8	2	5	6	4	3
		5			3	9		1
3		8		1	4	2	5	7
9			3	8		5		2
	7	1	2	9			3	
8	3	2		5				9

MEDIUM - 226

4		3		9			1	6
9	7	1		5			4	
8		2			7		3	5
5				8				3
2		7	6	1	4		5	9
6	9			2	3			
		5	9	6		3	8	
		6	4	7	1			2
7	2	9			5	4		1

MEDIUM - 227

			8	5	9	3		2
5	2	8	7	4	3	6	9	1
	7	9	1	6	2		4	
	5	7			4	2	6	
8						9	5	
								7
7		5	2	3			1	
	1	4	5		6			9
6	8		4		1	7	2	

MEDIUM - 228

8	4	5	1	6	7	2	3	9
9	1			3				6
3	2			5	9	8		
			9		3			4
	3					1	8	2
6	7	8	2	4		5		3
7	8		3				2	
1			5	2	4	7	6	8
		2	7	9		3		

MEDIUM - 229

4	5	9	6		1	8	3	7
		1	9	3	7		5	
7			5					6
	4	6				5		1
5				6	9			
9	3	7	2	1	5	6	4	
1	9		7	5		3	6	4
3		4			6	7	2	5
6	7		3	4	2	1	8	

MEDIUM - 230

5	2	3	4	7	6	8	9	
8	9	7	1	5				4
1					2	5	7	3
			7	3	1	9		
3	6		5				4	
9	7	1			8	2	3	5
7	8	9	2		4		5	
6		5	9	8	7		1	2
		2	3			7	8	

MEDIUM - 231

1	3	7	9	4		5	6	
9		2	1	5	6			
5		4	7				9	
2	9				5	3		
3	7	6	2	9	1	4		
4	5	8			7	2		9
	1		8	7	2		4	5
			6					7
7	2			1	4		8	

MEDIUM - 232

	4	6	2	7	9	5	1	
				4	8	9		
2	8	9	6			7	4	3
6		3			4	8	7	
9	2	4	8			6	5	1
		7		1	6		3	
7	9	2	4				8	5
		8		9		3		
1	3	5	7		2	4		

MEDIUM - 233

	9		2	4		3	8	6
	3				8	4	5	7
5	4	8	6	3			9	2
		2	9			5	6	4
9	1	4	5				7	3
			7		4		1	
1	8	9		2				
			3				4	
	5	3	8		9	6	2	

MEDIUM - 234

	8		2	7		9	4	3
	2	3	9					
9	4	7			6	1	8	
2			3	1			5	
	5	8		2	9	4		
7			4	6	5		2	9
	6					3	7	4
		2	6	9	3	5		8
	1	5	8	4		2	9	6

MEDIUM - 235

7	1	5			6	9	3	
	8	9	3		5			4
3			1				6	5
2	3	8	9				5	1
	7	4	2		1		8	
1	5		4		8	2	7	9
5	9	7	6			1	2	
		1	5		3	6		
4		3	7		2			8

MEDIUM - 236

		8		5			3	
3		9			7		1	8
				1	3		4	9
	9					3		
	6		5	8	2	4		1
4				9	6		5	
2	3		1	7	8	9	6	5
9	8	6	4	2	5	1		3
1			6	3		2	8	4

MEDIUM - 237

1							5	9
5				7	3	4		
3		2		9	8	7	6	1
2		3		5				7
7		1	8	3			4	
4			6	1		2	3	8
		5	9	2			7	3
8		7			5	1		4
9		4	7	8		5	2	6

MEDIUM - 238

	8		7	6			4	
3	5	2		1				6
	7	4	9	2		8		3
		8	2	7		1		5
	6	7		8			2	9
2		9	3	5		4	8	7
			5		7	6	9	8
		6			8	2		
	4	5	6	9				

MEDIUM - 239

		8	9	7	3	4		
1	7		5	8	4	9	6	3
	9	3	2		1			
9					8	3	1	5
3	1							9
8	6		3	1			4	7
7	3	1	4	5	6		9	
	4		8		2			
2			1		7	5	3	

MEDIUM - 240

	3			2			8	9
	8	9	3	1	6	7	4	2
		4	9	7		3		5
		7	2				9	3
	4			5	3			1
3		6	8		9		2	7
4		8					5	6
6	7			9	2		3	8
1	2		6	8		9	7	4

MEDIUM - 241

7	1	8	6	2		9		4
6			9			3	1	
		9	7	1	5	2		8
1		7		6	9	8		2
5		3	8			4	9	
		2	4		1	5		6
		6		9				3
	7	4				6		9
2			3			7	8	

MEDIUM - 242

		8	1	5		7		
	9	3				8		
	5	7	3	9	8		2	
				2	1	3		7
	8		9			6		5
3		1			5	9	4	2
5	3	9	7		2	4	6	8
	1	6		3	4		7	9
7	2		8					1

MEDIUM - 243

	6	7		2	1	5		
		8	6	4	5			
2			7		3		1	8
8	2				9	1		5
	9		3	8		2	7	4
7	4	6		5	2	8	3	9
3	7		2	6	8	4		
	1							2
4			9		7			

MEDIUM - 244

	5	2	8	4		7		9
	7	4	6	9	3	2		1
			7	5	2		6	
	9	7	2	1	5	6		3
5	2		9	3		4	1	7
3	6		4		8			
	8					9	7	5
9								6
	1	6	5	2	9	3	4	

MEDIUM - 245

		7	8	2				
	5	9	1					
			6	5	9		7	
7		1		9	8	5	3	4
5	2		7		6		9	8
9	4	8		1	5	2	6	
	7			6			8	
	9	5	4	3	2		1	
1	3	6			7		4	2

MEDIUM - 246

					7	6	5	8
3			5		1		9	7
5			8			4		1
9	1		2		8			
		3			5		1	
			1	9	3	7		2
7	9		3	8	2		6	
8	5	2		1	9	3	7	4
	3	6	7			8		

MEDIUM - 247

	9				4			8
6	2		9		8	4		3
	8			1	3		5	9
3	4		2	5		9	8	6
2		1	7		9			
8	5	9	3				2	
4	3		1		7	5		
9	1	2	8	3	5	6		
5		6			2	8	3	1

MEDIUM - 248

9	2		7	4		8		3
			8	3	9	7	1	
7	3	8	1			4		
		5		8	1	6		9
	9	2	5	6	4	3		1
	4	6			3	2		
					2			
	1			5	7	9		6
	6	7			8		2	4

MEDIUM - 249

6				3	5			
2		5	7	6	4	9	3	
	3			8	2	7	5	
7	6				3	5		
3		8				4	1	
9	5		8	7				
	7	6	3	9		2	4	
	4	3	2	1	7			8
	9	2	5	4	6	1	7	

MEDIUM - 250

	4	9	5	1	7	8	3	
7	5	8	2	9			4	
2	3	1	6	8	4		7	5
3			7					1
4			8	6	1	5	2	
							8	7
	6	5		7	9	2		4
		4	1	2				
1	2	3	4		8			

MEDIUM - 251

	6		1	9	4		2	
3		1	6	2	7	8		
	2		5			6	1	4
	9				3	1	8	
8	1	6	4	5		3	7	2
		3	8	1	6	9	4	5
1	8	2					3	
6						2		8
	3	4		8	2			

MEDIUM - 252

5	1	7	8			6	4	9
	2		5	9	4			
	9	3	7	6		5		
9	4	5		3		8	6	
	3	6	4	8	7			5
8		1				3	2	4
3				4	5		7	
						4		
7	6	4		1	8	2		3

MEDIUM - 253

6	1	4	9	7		5	8	3
2	8			3	5			
	9	5				1	7	
7		9						8
1		6	2	9				7
8	2	3	4	1		9		
	6			2	3	7	9	
			7	4	9	8	6	
	7	8	5	6			2	4

MEDIUM - 254

1	2	7		6	3		4	
6	4	8	9				5	7
3				7	8	6	2	1
	7		6		1	4	3	2
	6			5		1	7	
			2	3				
	8		3		9	7	1	
	9		7			2	6	3
	3	2	1		6		9	

MEDIUM - 255

	6			1	4	2		8
					9			6
	9	4	2			7	5	
6	4	9	1	5	8			7
7	2	8		4		9	1	5
5	3	1	9	2	7	6		4
	7	6		9	5		3	1
	8	5	7	3				
				6				

MEDIUM - 256

		1	6		9	8	5	7
	8	9	3	1	5	4	6	2
6		2			7			9
	6	7		9	2		4	8
8	2		5				1	
9	1			4	6		2	3
			2	3				1
2			9	5		3	8	
	4			6		2	9	

MEDIUM - 257

			8			7		1
		2			7		8	4
	7		1	5	4		9	
	6	5	4	3			2	7
	9					3	1	
3		7	9			6	4	
5				8			7	
7	3	9			1	8		6
8	4			9	5	1	3	2

MEDIUM - 258

8				2	7	9	1	5
7	9	2			5			
	1			8				2
		8				2		
		7	5	6		8	4	9
		1	8		2	3		7
5	8		2	7	6			4
		4					2	3
2	7	9	1		3	5	8	6

MEDIUM - 259

					3	1	8	9
	9		2		8	6	7	4
8	6	4		1			5	3
1			5		6	8	3	7
9	8		1	3	2	5	4	6
5					4		1	
4	5	8			9			1
			6	4				
6	2	3	7			4	9	

MEDIUM - 260

	5			1			3	2
7		3				4	8	1
4			7	8				9
	4	7				8	2	
3	2		8	7		1	4	
			2	4		9		3
	3	4		5		6		7
1	9		4	2		3	5	
5	7	8	3	6	1			4

MEDIUM - 261

8						3		
3			4	9	8		7	
					6	4	8	1
7	4			6	5			
9			8	3			6	7
1	3	6			2		4	8
5	8		6	4	3	7	9	2
6	2	7		1		8		
4	9	3	2		7		1	

MEDIUM - 262

1		9				6		
		7	8				1	2
2	4	5	1		6		3	
	2	6	4	5		1		7
7	5			6	9	8		4
4			2		1		5	6
	1	2	6	8	4	5		3
8	7		5	1	2			
	6				7	2	8	

MEDIUM - 263

7	3		1		6	4	8	2
8	1		3					5
					7	1	6	3
5			2		4		1	9
9	7		6		5			4
4		8	7	1	9		5	
3			4	2		6	9	
6		2	9	7	8			
1	9	7				2	4	8

MEDIUM - 264

7	9	6		1	3	4	8	
3	1	5			4	2		7
		8	5	7		3		6
5	2				1	8		
1					5	7	2	
		7	6	4			5	3
6	3	1	4	9		5	7	2
8		2			6	9		1
4	5	9	1	2		6	3	

MEDIUM - 265

	6	9	5				7	
	4	7	2	3		6		9
8	3	2	7	9		1	4	5
7			4		2		3	6
	2	4	3	6	7	5		8
3			1		9			
	5				3	8	6	
2		8						3
		3	8	7				1

MEDIUM - 266

	2		7				5	
		9		4		1	7	
6	7		5	1	3	8		
	5	8	4			3	2	9
		1	3	9				
3				7			1	4
2	3	6	1	5			4	8
8				3		2		5
9	4			2		7	3	1

MEDIUM - 267

		7			3	1	4	
	1	2	4		5	8		
8		4		9	2		7	5
2	6		8	5	7	3	1	
5	7	3			4			6
4		1	2		6	7	5	9
7			3			4	6	
3	4				1		2	
		6	7		9	5	3	8

MEDIUM - 268

4				9	2		7	
5		7	4		1	6	3	
	6	1	7				8	9
		6		4	7		5	3
9		4	2		3			7
7		3					4	6
8	7			5	6		9	
6		9	1					5
	1	5		2	4		6	8

MEDIUM - 269

			1	5		2		
2				9	3		5	4
5	8	4	2			3	9	1
1	3	6			5	9	4	2
8	2	5	6	4		7	1	3
7			3		1			
	6	1		3		5		7
9		8						6
			4	6	7	1		9

MEDIUM - 270

		5	2	4		6	7	
			5	3		9		
1		8						3
3		2	6				9	
		4	1		2	7		
6		1	7	8	3	5	4	2
5	2	3	4	1		8	6	9
			3	2	9	4	1	
4			8		5			7

MEDIUM - 271

2	3	4	7	9	8	5	1	6
7	1	8		6		2		3
			1	3	2	4		
	5				3			
	7	1		8		9	5	2
4	8					3		1
6			8	4	9			
		5	3	7		6		
1	9	3	5		6	7	8	4

MEDIUM - 272

1	5		4	8			2	
8	2	4	1					5
9	7	6			2	8	4	
2	1	7	8			5	6	
4		8	2		6		3	7
6	3		9		1	2		4
3	4		6				1	
7		9	3		4			2
5		1	7		8	4	9	

MEDIUM - 273

1		2	3	7		6	9	
7		3	6	8	4	5	2	1
6	5	8		2				7
	7				1	2	6	
	1	6			2			3
5			7			1		
4		7	5			9	3	2
2		1					5	
			2	3	8	7	1	4

MEDIUM - 274

8	3	7		1	5		2	6
				7	2	9		5
2		5		4	3		7	
9	7		4	2		5		8
	2	6		9			4	
	4		3		6	7	9	2
			5				8	4
4	8		1			3		9
	5	9			4			7

MEDIUM - 275

5	9			2		3		
	2	3				4		9
7	6		9		3			
		2	8	3	9	1		
9	4	7	2	6			8	
3		1	5	4	7	2	9	
8		5	4	9		6		1
		6	3	7	5	9		
	3				6	7	2	5

MEDIUM - 276

9	7	4	3	2			5	
6	5		4		9			1
	3		5		6	9	2	
			8	5	7	1		
7		1	9	3		8	6	5
8	4	5	1	6	2	7	3	
	6	7		1				3
2		9				5	8	
4		3	7					6

MEDIUM - 277

		4		8			9	
1	3		6	9				4
5	9	7		2	1			8
		9	5			2	1	3
	6	5	8		2			9
2	7	1			4		8	5
		2		7	8		5	6
	5			4	9	8		
9	8	3		5		1	4	

MEDIUM - 278

8	9	3	6	5	2			7
4	7				1	5	6	9
1	6	5	9	7			3	
3		8	4		9			
6	2		7	1		4		3
5		7		3	8	9	1	6
					7			1
			1	9		6	5	
9		1		2	6		7	

MEDIUM - 279

				8	3		7	2
3	9			6	7		5	
8	7		5		4	3		
				5	1		4	6
1	3		4	7		5	2	
7	5	4	6	2			3	
		9	7				1	5
5	1			4	6	2	9	7
2	6		1	9		4	8	

MEDIUM - 280

4	5	7		3		6	2	8
9	1		6		8			
	8	2				3	9	1
1		5						9
2		8	5	6			1	7
7	9	4	3	1	2	8		
		1		8	3			6
				5			7	
	7	6	1	9	4			2

MEDIUM - 281

4		7	2		9			8
	6			5	3	7		1
1	2		8		7	4		5
9		8		2	5			4
5	4				1		6	7
6	3		7	4	8	9	5	
2		6	5	7		1		3
		4	1		2			6
7				8				

MEDIUM - 282

8	4		2	5		9	7	
2	6	7		1				4
		5	7	8				
		8	4	9		7		5
	2		1	7		4		9
9	7		6	2			3	
			5	3	7	1	9	2
	3	2		4	1			8
1		9	8	6				7

MEDIUM - 283

2		7				4	3	
			3		7			2
3	4	8	5		9		6	
8	2	4		6	1	5		3
9	5		4	3	8	2		
7		3					1	
4	8		6			7	5	9
1		6	9	5		3		8
	3		8		4	6	2	1

MEDIUM - 284

			3				1	
2		1	6	9		3		
8	4		2		1	6	9	
7		6			2	5	4	
		4	7				8	3
	1		4		5	2	6	
4		5			6			1
3	9	2	8	1	7	4	5	
1		7		4	3		2	

MEDIUM - 285

8	3	7		2	1		4	
9				6		2		
2	6		3	5	7	8		
			7	1	6			
1	8		4	9	3	7	2	5
4	7		5	8	2	6	1	9
3		2	6	7	5	1		4
6							5	7
		8		4	9	3	6	

MEDIUM - 286

			2	4		8		5
		5	7	3				4
4			6	9		3		7
			5	7		6	4	9
6	9	4	8		2	7		
3	5	7					8	1
5		2	3	8			9	
7		9		5			3	
	3	6	9		4	5		

MEDIUM - 287

2			4		6			5
9		3	5	7	1		4	
		5	8	2		6		9
1		2	3	6	5			4
6	3	4	7	8	9			2
		7	2		4	9	6	3
3			6			4		
	4		1	3		5		7
			9	4	8	3	2	

MEDIUM - 288

2			9		4		6	1
8	1	9	5	2	6	4		
6	4	5						
4						2	7	9
1		2						8
		8	2	6	9	3	1	4
	7	6	4		2		8	
			6	7	3			5
3	2		8		5		4	6

MEDIUM - 289

2	8	3		5				7
7				3		6	5	
			2					3
	2					1	9	8
8		9	3	2		7	6	4
4	7	1	8	9		2		
5				8	9	3	7	6
		7	5	6			4	
9		8	4	7	3	5	2	1

MEDIUM - 290

4			9	8				1
9		1			2			7
6			7		1	8		
	1	9	2	5			7	8
2	7	8	3	9		4	1	
	6			1				
8			1	4	3	7	2	
7			5			1		3
1	2		6	7	8	9	5	

MEDIUM - 291

		1		8	3		6	
7	4	6		1	5	9		
5		8				1	4	
2	8			4	7	6	5	
4		5	8	6	2		1	
	1					8	2	4
			6	9	8		7	
3		9		2		5		
8		4	5	3	1	2	9	

MEDIUM - 292

					6	2	7	5
7	3	2	4		5		1	8
6		8				4	9	
3		5	6		9	8	2	4
				5	2	9	3	
1	2	9	3		8			6
5	6			7				
	9	3		8	1	5		
2	8		5	6	3			9

MEDIUM - 293

3				7		2		1
	1	7	2			3	5	
	5	2			1	7	6	
	9			5	8			2
8			6	1	2		4	7
6	2		4	9		8		
5				2		1	7	3
	6		7			4	2	9
		9	1	4		5		6

MEDIUM - 294

		7	8	1	9		4	
2	3		6				5	8
	8	9				6		
	7	2			1			
6		4	7	9	2	8	3	
	9			6		2		7
9		3				7	2	4
1	2		3	4	7	5	8	
7	4	8	9	2	5	1	6	3

MEDIUM - 295

8			9		7	6		
		3	6				9	7
7	9	6		3			2	1
2		5	1		3			4
9			5	2	6	1	3	
		8	4	7		2		5
1		9	2	4	8	7	5	
4	8	2	7	6	5			9
6		7		9		4	8	

MEDIUM - 296

	1		4	2		8	6	3
6			3		7	4		5
3	4	2	6		5	7		
8	9		5			2		4
	5	3		9	4			1
1	6	4	2		8	9		7
9	3				2			6
			9	7	3			8
	7	1			6		9	

MEDIUM - 297

4	5			1		6		8
7	1	2		8				
	6			3	7	1		2
3	2			5	4		9	7
	4		7		8	2		
8		1	3				6	5
2	9	4		7				1
1	8	7		4			5	
6		5				7	2	4

MEDIUM - 298

1	2	8				4	7	
5		7		6	4		9	1
9	4	6	2	7	1		5	
4	9			1		7	6	
	1	3		9	6		2	
	6	5		3		1		
2		1				8	3	
6	8	4		2		9	1	
	7		1		5	6		

MEDIUM - 299

1	5		9		6	2		8
8	9		5	2		3		
	3		1	8	4			9
9	7						1	
	6		3	5	1	8		
5	1	3	8					
6	8	1		4		9		
3	4	9		1	8	7	6	
7	2	5	6			1	8	

MEDIUM - 300

		7	2	5		9		3
6	9	3	4	7		8		
8		5	9	3	6	7	1	
	5	6		2		3		8
7		8		4	3	5	2	
2	3	9		8	5		4	1
	6	1		9				
9		2					8	5
3	8		5	1	7			

MEDIUM - 301

2			8	6		3	4	1
	1	3	7	9	4	5		2
	5	4	2	3	1		9	7
5		8				1		9
	6		1	8		4		
	3		4	5	9			
			9	4	7	2	5	6
	2		5		3	9		
4			6	2	8			

MEDIUM - 302

7		5	4			6		9
		1						8
	2		6	8		7		
8		6	9		7	2	5	1
1	3	2	8		5	4	9	7
9		7	1	2		8		
		9	3	4	1	5	8	
		8		5	6			2
		3	2	9		1	7	

MEDIUM - 303

	6	7	9	3		5		
9			1	8			2	
8		1	4	5		6	9	
5	1	4			9	3		6
6			3		1		5	4
		2			4	8	1	
3	4	8			5			
			2		8		3	5
2	9	5			3	7		

MEDIUM - 304

1		3	4		9		8	
	6		8	7	3	4		
		9	5	6	1	3		
7		1		9			3	8
	9					2	7	4
4		8			7	1	5	
5			7	1		9	6	3
			3	8		7	4	5
6	3			4			1	

MEDIUM - 305

			5	6	1		2	8
	5	6	7	8		3	4	9
7	2			4	9	1		
9	1		2	3				6
	6	3				9	8	
8		5		9				4
5	9	7	6				1	
6	3	1		7				2
	8			1				7

MEDIUM - 306

8		6	7			5		3
			3	5			9	6
		4			8			
9	1	8	4		2			5
		2	5			4		
4	7	5			3	2	6	9
6				1	7			4
1				3		9	5	8
2	8	3	9	4		6	1	7

MEDIUM - 307

		8	2	9	1		4	
7		4	8		6	9	5	1
1	3	9	4	7		2		8
	7						1	
3			1	5	8	4	7	6
8	4		9		7	5		2
			5	8	3		9	
9		3	6	4		7		5
4				1	9			3

MEDIUM - 308

		1	3	4	6	2	8	
3	4	6		2			5	
8	2	9		1		4	3	6
7		4	6	3		5		2
9	5					6		
1		2	4		5	8	7	
		8	7		3		2	5
6			2		4	3	1	
2						7		4

MEDIUM - 309

8		7	1		5			2
3		6	9		4	5		
	1	5	3				6	4
6	5	1	4	9		2	8	7
	3			2	7		5	
7	8		6	5			3	
2			5		6		7	3
1	6			3	8			
5			2	1				6

MEDIUM - 310

5	2							9
9					4			
	4		8	3				1
	5	7	3	6	2			
6	3	2		9				4
8	9					2	3	6
	7	5		4		8	1	3
1	6	4	2	8	3	9	7	5
			7	1	5	6		2

MEDIUM - 311

8			6		5		2	
		5			3	1	4	
7	1	3	2					
3	5	8	9	4	2	7	1	6
9		1	3	6				2
4					7	8		
	8		5				6	
5	3	6	4		1	9		
2	9	4		8				1

MEDIUM - 312

	8		6	2	5	9	3	
					9		8	7
9	1	5	8	3		2	6	4
		6	7	1	8	4	9	2
2	9			5				8
7					3	1	5	
			3					5
	3	2		7		6	4	9
	7	4	9	6		8		3

MEDIUM - 313

				2				
		6	1	9				2
8		2		4	6	3	1	9
	2				4	7	8	1
6	3	4	7		1			5
		8					3	6
1	8	9	4	7	5	6	2	3
4				1	3	9	5	8
2	5	3			9			

MEDIUM - 314

				5	6			4
5			8			6	1	
1				9	2	8	5	7
	7	5				3	4	
6	9	3	2	4	8			1
8	1		5	3		2	9	6
3	5		4	6	1			9
	6					1	2	
	8	1	7		3		6	

MEDIUM - 315

			8	2	1	5		
8	6	1	4				9	
5	4	2		3			8	
2		5		8	4			
4	3	7	5			8	1	
6	8	9			3			
9	7		3	5	8	4		
1	5	8	2		6	9	7	3
3	2	4		9	7		5	8

MEDIUM - 316

	6		9	4		5		
5	2		1					
9	4	3		7		6	8	
		2	3	9				
		5	4			9	3	
	8			6				
1	3	6	2	5	7	4	9	
	9	8	6		4	7	1	5
7	5	4		1	9		6	3

MEDIUM - 317

1			8					
	7		3	1	4			
3			5	9		6		4
	3	4	2	7	9	1	6	5
	5	1	6	4	8		7	3
9			1	3	5	8	4	
5	8					4		
	1			5	2		3	8
7				8	1		2	

MEDIUM - 318

8	3			2	1			6
6	7	2	8		9			
		9	3		5	2	7	
2	5	4		1	8	6	3	7
	6	3		5			8	
		8		3	4	5	9	2
	8					7		
3	2	7		8		4		
	9	6		7			2	5

MEDIUM - 319

		5		7		8	3	2
3				4	1	6	7	
			8	3	2	4	5	1
7	1		3	9	4	5	2	8
9		2	6		5	7		4
	5					9	6	3
4				5				7
2	8	1	7		9			5
	7	9		1		2		

MEDIUM - 320

	8	5		3			1	9
4		9	1	7	5		6	2
6				4	9	7	5	
		4	6		2		3	7
		2		8	3	5		6
3		6			4			1
1		8		6	7			5
	6	7		2			9	
2			4		1		7	8

MEDIUM - 321

	1		4	9	6	5		
	9	4	5		3		1	
	3	7	2	8			9	
4		9	3	1		7	5	6
	8	3			7		2	
7	6	5	9		4		8	1
9		6			5	8	4	
		1	8	6	2	9		
		2		4			6	3

MEDIUM - 322

			5		7	4	9	8
5	8	4	9	3				2
9	7				6		3	1
8		5		4	9			
	9	6	8	1	2	3		5
	2			5	3		4	
6		8	2	7			5	3
	4						8	
7	5			9	8	6	2	4

MEDIUM - 323

	5	7	8	4		1		6
1		6	3	5	9		8	
9	4	8		6	7	2	5	3
4		2		9		3		
7		5	4		6		2	
		9		8	5	7		
8		3				9	7	1
	9		6				4	
5					8			2

MEDIUM - 324

8			1	6		3	9	
3	2			7	4	5		
					3	7	6	
	7	8	6	2				
9	3	2	4	5	7	1	8	
6		4	3			2	5	
4	8	5	7	1		6	3	
	6			3				
7	9	3			6	8	2	

MEDIUM - 325

3				7	1		6	
	4	6			2	7		5
			8	6				1
		4	2	9		8	1	3
7	2			8	3	6	5	9
	8				6			2
8	5	2					3	7
4			7			5	2	6
6	1	7	3	2	5	9	4	8

MEDIUM - 326

9	7	3				8		
6		5			2			7
		2	7	9	8		6	
2	4		1	7	9			
3	6	7			5		9	4
1			6	3			7	
5	9	4		8	7		1	6
7		1		6			8	9
8			9	4		7	5	2

MEDIUM - 327

1	2	4				6	8	9
			9		8		4	2
9			2		1			5
8	4				3	5	1	
7		5	1	8	6	4	2	3
6		3	5		4			7
		9	8		7			
5			6			9		4
2	7	1	4	5			6	

MEDIUM - 328

8		1	4		6		3	5
	6				1	2	4	8
5		4	2				1	
		7	6		5		2	
2	8			4	9		6	
1	9	6		2	7	4		3
3		9	7	6		5		
6	5		9			3	7	
7		8			3	6	9	2

MEDIUM - 329

			3	8				1
		8	1		5	7	9	6
	1			7		3	8	4
	5		4		1	8		9
		2		9	6	1	3	5
1		3	8				4	7
3	2				7	9		8
5			9	6		4	7	2
	7	4		1		5	6	

MEDIUM - 330

2	5				4			
	4			6		9	1	8
6								4
7		5	6	2	8	1	4	
8	6	4	3		9			
9	2	1	4	7	5	6	8	
4	9			5			7	1
		2	7	9	1		3	6
1	7	6	2	4	3	8		5

MEDIUM - 331

6	3		8	9				
	4			2	1	8	5	6
8	1		6	4	5	7	3	9
9	2		1		8	6		
			4	7	6		2	
5	6			3			7	
1	9			8	4			
4	7	6	5				8	2
	5		7			4	9	1

MEDIUM - 332

4		3	2	5				
6	5	9	3		1	4		
1	8		4	9	6	7	5	3
2		4		8		3	7	1
3	1	8	7	2		9		
5						8	4	2
7		1	9		2			6
					7		3	
8	3	6	1		5			

MEDIUM - 333

	1	7						4
9		2				7	6	5
3	4			2		1		9
7	5	3		9		4		
	9	4		6	1	5		2
1	2			5	7	8	9	3
	6		5	7	3	9		
4	3		6			2		7
5				4			3	8

MEDIUM - 334

2	7	5	3			4		
8	4	6	9	1		3		7
	9						5	8
			7	3	1	9		5
			4			1	8	
6	1			8	9	7	4	
	3	4		9			1	6
	8	9	6		3		7	
5		2	1	7			3	9

MEDIUM - 335

1		2			3	5		
5		8	2	6	4	7	1	
4	9	7		5		6		
7	8						5	1
			9	8		3	7	4
9	4	5		3	7		2	6
8		9		7	2	4	6	
		6	5	4		1	9	
	5		6		9		8	

MEDIUM - 336

	4		2			8	5	7
	9		4	8	5	1		6
5		8	1		7		9	4
	3	4					1	9
8	5			7	1		6	
2		6					8	3
	6		8	1			7	5
1			5		2		4	8
9		5	7	4	6	2	3	1

MEDIUM - 337

1	4	7				3		8
3	2	9		1	4		7	6
		5				1	2	4
9		8			1	6		2
7	5					8		
	1	6		8	5	7		
5	9	1		7		2		3
	7		6	9	3	4		
	6	3	1			9		7

MEDIUM - 338

	6	9	5					
		1	3	2	9	6	8	5
3	5	8	6	4			1	9
8	1	4	2	5		9	7	
		2			8		4	
6	3		7		4		2	
1		6	4		5	8		7
		3			2	4	6	
		7	9	1			5	

MEDIUM - 339

	6	8			4		7	5
		5			9	4		8
	4				8	3		1
4	8	1		2		6		
		7		3	1	2	5	4
	3				6			
9	7	4		6				
	1	3	4	8		7	9	6
8			7	9	3	1		2

MEDIUM - 340

		7					2	4
			3	6		7		1
					1	6	9	
8			5	4	7	2		3
7	3	4				5	6	9
				9			8	7
1		9	2	3		8		
	7	8	4			1		6
3	6	5	1		8	9	4	2

MEDIUM - 341

1			4	6		3	8	
4	8	5		3		6		
	3				9	2	4	5
		3	8	7		1	5	4
6	1	7		2	4		9	
5		8		1	3		6	
	6	4	3	5		9	7	
	7	2		4	1		3	
3	5	1			8	4	2	

MEDIUM - 342

3			1	7	5	6	9	
	4	7	3			5	2	1
		6	2		4	7		8
6					2	8	4	
2	7	4	6		8	9	1	
		1			7		6	2
	6				9	1	5	
5		3	8		1	4		9
		9	4	5	3		8	

MEDIUM - 343

1	2	5						8
					9			
	6	9	4	5	8	1		
	1				3	8	6	5
6	5		1		4		3	9
		2	5				4	1
3	4	8	6	7		9	1	2
5		1	9			6	8	4
			8	4	1			7

MEDIUM - 344

		8	9		1			
7	5			3		8	9	2
9	3	4	2	8	7	6		5
	1				9	5	2	8
		2	5	1		7		
8				2	6			9
	8			9	5	3	4	6
1					3	2	5	
5			7	4		9	8	

MEDIUM - 345

	1	2	6	4	7	9	8	5
	9		2		5		7	3
				3	8	4		
9						7		
	8	7	3		4	1	5	
5		1	7		2		9	8
4	2		1			5		
		8	4	7	3	2	6	
7	3	9	5	2			4	1

MEDIUM - 346

			2				8	7
		4	7	3		1		9
7	9			8		2		
	8	1	5			9	3	
	6	7	9	2	3	8	1	5
	5			1	8	7		4
	3		1	9		6	4	2
9	4	2	8				7	
1	7			4			9	8

MEDIUM - 347

		8						1
6	1			2		4		3
	9		8	1		7		
4	7		6	8		5		2
	6	2		7		8	1	9
	5	1	3	9	2			4
9		7				1	5	
1	8	6		4		3		7
5	2	3				9	4	

MEDIUM - 348

9	1	2	4	7	3	8	5	6
4	8	7		6		2	1	
	3			2				
				9				7
		9		5	4	3	6	2
	5			3	8	1	9	
	6		2			9	3	
7	9	8		1				
3	2	1	5		9		7	8

MEDIUM - 349

	7		9			2	3	4
		9	3			6	7	
			6		7		5	9
	1	8	5		3		2	6
7	2	5			9	4	1	
	6		2	4	1	7		5
		2		3			4	8
8	5					3	6	7
6		7	4	8		1		

MEDIUM - 350

6	9		8			5		3
4		7				9	2	8
3	5			9	2	6	7	1
8	4	5	1					9
	7	3		5	6		1	4
1	2		9				5	7
	3	4				1		
	6	9	5	1	3	4		
		1	6	4		7		

MEDIUM - 351

	5		7	6		3		9
7	9		5	2			6	
			9	8	3	2	7	
	7	2	3	5	9	8		1
9			8	7		5		2
8			4	1			9	
3	6	7		9				4
	8	9	1	3		6		7
		1		4		9		3

MEDIUM - 352

	7	2	6	1		4	3	
4	8			5		1	7	2
			7			6	8	9
	9	4	3	8		2		7
1				9		8		
	3	7	2		5	9	1	
	5	8	1	3		7	9	6
	4			7	6			1
	6		5		9		4	8

MEDIUM - 353

6	9	1					2	
8	2		6				3	7
4	3	7		5	2	6		1
5	8	4	3	6		9	1	2
1		2	5		4	3		
	7	3		8				
		6	7				5	
		8	9	2	5	7	4	
				4		2	8	

MEDIUM - 354

	7	2	1	5				6
6					2		5	
1		4					2	
5	1		9	7			6	
4	2		5	3	1	7	9	
		7	2	6		3	1	5
3	4	1	8	2	6		7	9
		8	3	1			4	
2	6	5		9	7			3

MEDIUM - 355

	6		1				8	
		1		8		4		2
3	7	8		2			1	9
	4	9		3	8			6
6		5			9	8	2	4
8	1			6	4	3	9	
	2	7		5		6	3	8
9					3	2		
1	8	3		7				5

MEDIUM - 356

1	3	7	5	6	4			
	9		7		3	1	5	
5			9	1	8	3	7	
7		3	1	5	2	6		8
2	8	6		7	9		1	3
	5	1			6			
4		5				8	3	
8								9
3		9	6	8			4	

MEDIUM - 357

6		1				8	2	
2		5	8					9
	7	4	5	2			3	1
7	8	6	9					3
	1		7			2	8	
			1		8	9	6	7
5		3	2	8		1		
1		8	3	9	4	7	5	2
9	2	7		5				8

MEDIUM - 358

5	1	9	8	2	4			6
2	8		5	7	6	4	9	1
6	4	7				2		
					5	3	6	8
8							5	
4		6	3	1			7	2
9		5	4	8				3
		8		6		5	4	
				5	2	8	1	9

MEDIUM - 359

	1							
2	8	5		6		1	3	9
9	3		2			7		6
5				4	7		6	
		2	6	8		5		
1	7	6	5	2		4	9	8
4		1	3	7	2			5
	2	9		5		6		3
			8		6	2		

MEDIUM - 360

5	2	1		3	9	4		7
7	3	9	2	6	4	1	8	5
	6	4	7	5		2		
9		3						4
	7				5			
6	1		4	9		5		8
	9	6	5	1	8	7	4	
1			3		2	9	5	6
	4				6	8	3	1

MEDIUM - 361

		1	3	9				2
	9	2	7		5		8	3
	7	4	2	6	8	1		
1	3	7	5	8			2	4
8	5	9		3		7	1	
	2	6			1			8
7		5				2	3	
2	4	3		5	7	8		
				2		5		

MEDIUM - 362

6							1	
5	9			7				
2	3	8	1	6	4	7		5
		3		5	1	9		
9	2	5	6			1	4	7
	4					8	5	
3	5	7		1	8	4	2	6
		9	2	3		5	8	1
8	1	2						9

MEDIUM - 363

2	3	8			7			
9	4	5		2	8		3	6
						4	8	
		9	2	5	6		1	
		2		3	4	5		
							2	3
	2	7	4			6	9	5
4	8	3		6		2	7	1
5	9	6	7		2	3	4	8

MEDIUM - 364

1	5				4	6	3	9
			9	1	5	7		2
2	8		7	6	3	1	5	4
	7	8	6		1		4	
		2			7		1	6
4	1	6	8			2		5
	2		4	7			9	1
9	6	1		5				
		4	1					8

MEDIUM - 365

	7	5	1	2				3
4	9				5	8	2	
1		2		4		6		7
2		6		3	7	1	9	5
8				5	6		3	
3			4	1		7	6	8
		8		7			1	2
7			2	9		5		6
		1		8		4	7	9

MEDIUM - 366

	9	2		1		7	5	
				3	9		2	8
8					5		6	
9	7	8		2	6	5		
1			8	5	7	9		2
2	3		4	9	1		7	
	2	1	9		3		8	7
4		3	1	7	2			5
7				8			1	3

MEDIUM - 367

	6	2		5	1		7	3
	1	7	3	6		2		
	8	3	7				1	4
	7		9					6
	4		1			3	2	9
	2	9		4	3	5	8	
7	5	8			4			1
6			8		7	4		2
	9	4	5	1	6		3	8

MEDIUM - 368

3			6					4
5		1		3				6
	6			1	5			3
8			7		6		2	5
6		5		8			4	9
4		7	3	5	9	6	8	1
1	5		8	7	3	9	6	2
	9	6	5	2				8
	8	3			4	5		7

MEDIUM - 369

3	1		5		8		2	
				4	3		1	6
5	2		6	9		3		7
4		1	9		7			
	5		4				9	1
9	3		8			6	5	4
1	8		7	6			4	
	4			8	9	2	7	5
7	9		3		4			8

MEDIUM - 370

	6		9	4		7		1
						8	9	4
8			1					
			7			1		6
9		7		6				3
	3	6	4		2	9	7	
7	2			8	4	3	6	9
	5	3	2	9	1	4		7
4	8	9		7	6	5	1	2

MEDIUM - 371

		9	4	3		7		
	8		5	9	7	3	4	
	4	7	8	6	1	5	9	2
5		4	7	8		1		
			9	1	6	4		
	1							
2	3	8		4		6	7	9
9			3			2	1	4
			6	2	9			5

MEDIUM - 372

2		5	6	8	3	9	1	7
7	1	8			9	3	6	
			2	7	1	8	5	4
		7	4			6		
4				3	8			9
8	5		9			4	7	1
5	7	2			4		3	6
1	8	4		6		2	9	
	9		1					

MEDIUM - 373

				8	4	1	6	7
	8		6	1	3	4		
4	1				2			
8	3	1			7	6	4	5
2	5		3			8		9
6			1	5	8		7	3
		9	2			3		
5	7	8	4	3	1	9		
3	4	2	8	6			5	1

MEDIUM - 374

	1		8			4	6	3
		3	9	2	6	8	1	
7				4	3	2		9
		7	4		2	5		
8		1	7		5		4	
		5		6			8	
	7	4	5	1		6		8
		8	2		4		3	
5	9	2	6	3		1		4

MEDIUM - 375

7	1		3		4	5	8	9
	3	4		8	5	1		
5		8		7	9		2	4
	4	7	5		6	2		3
	5	3	7					
			8	4			1	5
	2	1	9		8	6		7
	7				1		3	
	8	5	6				4	1

MEDIUM - 376

			9				2	4
	4	9		3	2	5	7	
3	2	5						9
2	7	1	4		3		6	
4		6		8	1	7		2
5	3	8	6				9	
7	1	2	3	4	6		5	
6		3	8					
9			2		5	6		

MEDIUM - 377

4		3	8	9		2	7	
	6	1		3	2		9	
2	8	9	1	7		6		
8			4		5	3	6	7
1	4		3	2	7	9	8	
3			6	8		1		4
6			2		1			9
9		4	7	6	8			
		7	9					6

MEDIUM - 378

7		8	9			3	6	5
6				7	5		9	
						1		7
9	4	7						3
2	6		3			7	5	4
	5	3	7	4	2			9
	7		6	3		9		
1	9			5	7	4		
3	8	2		9	1	5		6

MEDIUM - 379

		5	1	9		8	2	3
9	1			3	4	6		
	3	7	2		6		1	
		4	5	2			6	
5	8			7	1			9
2	7	6	4	8	9		3	1
4	2		3	6		1	8	
3		8		1		4		6
	6		9				5	2

MEDIUM - 380

	7			1	2		5	
2	3		8		7		4	6
1		9		5				7
5		3		2	8		6	4
7			4	3	6	5	1	
9	4	6	1	7				2
		1		8	9	4		5
8		7			3	6		1
3		4		6	1			

MEDIUM - 381

5	6		4			3		1
	7			5	6		4	8
4		2	8	9	3			
7			9				2	
	4		6	2		9	3	
2	9		3			7		
9		5	7		8	4		
6	3	7		4		8		9
1	8		5		9			2

MEDIUM - 382

		5			7		3	
	7	2	5			4	8	
8	9			1			5	7
	2	8			1	5		6
	4						9	1
			6	8	5			3
	3	7	4	2		9		5
	8		7	5			2	4
	5	4	1	3	9			8

MEDIUM - 383

6	9	3		7		4		
5			9	4				3
1	4	2	3	5	6		8	9
4	8	9	1	3	5	2	7	6
7				6	2	8		
	5	6	7				1	
				9		5	6	
			5		7	1	3	
8					3	9	4	

MEDIUM - 384

3		4	2	7	8	5		
1	7	5	3					
2	8	6	4	5	1	3		
	4	2			3	1		5
5	3		1			9		8
9		8	5		7			
7	6				2	8	5	4
				8	5		3	1
	5	3		1	4	7		

MEDIUM - 385

						8	5	1
	5	8	3	4		6	7	9
			5	9	8		4	2
5			4		3	1	9	8
	8		2			5	6	
	1	9	8		5		2	4
	4	5			6		1	7
6	7						3	5
9			7				8	

MEDIUM - 386

8	6	1		3	9			
	3		7		6	1		
7		4	1	5	2	8		
3	2					5		
1		7			5	9	3	6
9		8	3		7		1	2
6	1		5		8	2	4	
2		9			4	3		8
					3		7	

MEDIUM - 387

4		6	1	9	2	3		8
3		9		7		1	4	5
	7	1	5	4			6	2
	1			6	9			4
2		7	4			6	5	
6		4				7	3	1
9					6	8		
7		2	9	8	4	5	1	6
		8		5			9	3

MEDIUM - 388

2	6	3		9		7	4	1
	8	1		3		9	2	
	5				1		6	
	1		9	4	7			
3	7		5	6		2	1	
				1	2	4		8
1	4	6	7	8			3	
	2	7	1		3			4
		8		2	6		9	7

MEDIUM - 389

8	3	4		2			9	
6		5	8	7		2		3
9	2	7				5		8
2		3		1	7	8	5	
	9		3		2	4	7	1
1	7	8	6	5	4			9
			2		8		3	
3				6		9	8	4
7				4				2

MEDIUM - 390

6	3	4	9		8	2	1	7
2				7				5
5	8	7		1				4
	6	5				1		9
3			6	4	9	5		
	7	8	1	3		4		
	4			9	7		5	2
8	2	9				7	4	3
7	5		3			9	8	1

MEDIUM - 391

2	8		7	5		6		9
		6				5	7	
		5		6	1	3	8	
8	3			4	6		2	5
	2	7	8	3		9	4	1
1	5		2			8		3
		2		8			9	
3	9		6				5	
		1	5	7		2		8

MEDIUM - 392

4	2	9			1	8		7
1	7	8	4	2	6	9	3	5
6	5	3					1	2
		6		4	3			
		2				5		
5				9	8	6		3
3	1	5	8	7	4	2	9	6
8	6	4		3		7		1
	9	7		6	5	3	8	

MEDIUM - 393

9	5	1	4	8	6	7	2	3
	6	3	5	9	2	1	4	
4	8	2	7	3				5
2	7						1	9
	1			7	9			
	3	9	6	1	5			4
	4		9	6	7	3	5	
		5	1		4	6	8	7
	2				8	4		1

MEDIUM - 394

3	9	4			1			
5	8	7	4		9		6	1
2	1					4	5	
		1	6	3	2		9	
					5			
9				4	8			6
8	3	9	2			6	7	5
1	6	5		8	7		4	2
4	7	2	9		6	1	3	8

MEDIUM - 395

	1	3				7	2	
2					9		3	1
		9	1	3	2	8	4	
7	5	2	9	1				
9	6	4	3		7	1	5	
		1	2	5		6		7
		6	8	2				4
	8		4		1	2	6	5
				6	3	9		

MEDIUM - 396

	8	1	3		4		7	
5	4	7	1	2	6	8	3	9
3			9					
				9	5	7		1
7		2					9	
	1	9			8	2	4	
4	7	8	5		3		2	6
2	6		8	4		1		
		5	6	7	2	4	8	3

MEDIUM - 397

2	3	9		5	7	1	6	4
1	5		9	2	6	3	7	
8					3		5	9
		5	1		8		9	2
3	1		6		2			7
	9	2			4	8		3
	2	3			9		8	
9		6		4	1	7	2	
	7					9		6

MEDIUM - 398

3	5	6	2	4	8	9		7
		4				5		2
		9	3		5	6		4
5	9	8			3		7	6
		7		8	6	1	5	3
	3			2	7	8		
9	1	2		5		3	6	
				3		7	2	
8	7			6	2			5

MEDIUM - 399

5	1	7			8		3	4
	3	9		5		8	7	
8	4		3		7			
1	2	4	7		9	3		5
7	5	6			3	9	8	2
			5	2	6	4		7
		1	8			5	9	
9				6	1			3
4	6		9		5	1	2	8

MEDIUM - 400

	8			6		3	5	2
5	3	9	8			6		4
2	6		3	4		7	8	9
						4	9	3
	1	3	2	9		8		5
9	4		7	8		1		6
		7	6		9	5		8
	5	8	4	2		9	6	
4	9			1	8		3	7

MEDIUM - 401

8		4	6		7			1
	6			8	1	4	7	9
	2	1		5		6	3	
3		5			9		2	6
9	7				2		4	
6	4		7	1		8		
	3		1		8	5		7
2			5	9			1	
	5	6	4	7	3	9		2

MEDIUM - 402

7	5	9	2				8	1
				1	6			7
		1				4	5	
8	9		4		5	1	2	6
1		6	3			7	4	5
		4		2		8		3
	3	5	7	6				
2	1	7	9		4	5	6	8
6	4	8	1	5	2	3		9

MEDIUM - 403

	5	1	2		9	3		8
3	4				6	7		
2		6		7	3			
	1			8	2		5	3
	2	9			7		8	6
8			6	9				7
			1	6	5	8	3	4
5	8	3	7			6		
1		4	9		8	5		

MEDIUM - 404

4		9	1		5	7	2	
1			7			8		
7	2				3			
	7	8			1	3		2
2	5		8	3		1		6
3		1	2		6			
9	6		3		8	2		1
	1		9				6	7
5	4		6	1		9		

MEDIUM - 405

			6					
	9	8	2			5	6	4
	3		9		7		2	8
	5	7			6		9	1
8		2	4				5	7
		3	7		8	6	4	2
3	7	9	1	6	4	2	8	5
				3	9		1	6
6	8	1						

MEDIUM - 406

9		6		8	2	7	3	1
			3	5	6		8	9
3	2		1					6
		9	7		1			
2	3	7			5	6		
1			6	2	3	8		7
	1			6	8	4	2	5
	9	4	2				6	8
6					4	9	7	

MEDIUM - 407

7	5	4	6		2	9	3	
2	9		5	1			4	
8	3		4		7	5	2	6
	4	8	7	3	1		5	2
3		5			6	8	7	
6	7	2		5				
			8	2			9	
		7	1		9	3	8	
			3	7	4	2	6	5

MEDIUM - 408

	8		4	9			2	3
	4	3		7	2			9
	2	9		5		1	4	
	5	2	7			9	1	
9		1	3		4	5	6	8
	6	8		1		3		
	1	4		3		6	9	7
2		7			8	4		5
	3	5		4				

MEDIUM - 409

			1		5	4	3	8
			2				7	6
6	4		7			2		1
	6	7	8		9			3
9		5			7	6	8	2
			6	2	1	5		7
		6		8	2		1	
5	2	1	9	7				4
		9		1	6	3		

MEDIUM - 410

6	5	3			2		4	8
		4		8	7		3	5
	2	8	3	4		9	1	
4	6	9	1		8	5		2
	8	2	7	5	4	6	9	3
3	7		2				8	4
			5		1			
						8	5	
5	4		8		6	3	2	

MEDIUM - 411

6		5				8	4	1
4		8	7	5		3	2	
2	3	1		4	8		9	
			1	6	9	7	8	5
1	5		3					
8	6		5	2	4	9	1	3
7	4	6	8			2		
	1			7	5	6	3	8
5								4

MEDIUM - 412

8	4	5	1	2		7		
	9			7	3	8		6
7			8	9			1	
	1		6	8	7	2	5	3
6	2	8	3		5	9		
5	3				2			1
						1	2	
2		4		6				8
3	8	1	2	5				7

MEDIUM - 413

1		9	4	8	2		5	
	8		5					
	4	6	1	9	3		7	
6		4	9	3			8	2
		2		5	8		4	9
				4		5		3
4					9	8	3	
8	6			1	4	9	2	7
9		7	8		5	6	1	4

MEDIUM - 414

4		7	8	5	6			1
9	5	6	2		1	7	3	
8				3		6	5	
	4	1	6	2	3			9
	9	8	5		4	2		3
2		3		8	7			5
	7		1	9				2
3				6	2		9	7
1							4	6

MEDIUM - 415

3		1	9			7	4	2
9	2	6			7		5	
7	4		3	2	8			
	6	3	5		1			
4		8	2				1	
				7	6	2	8	3
5		7	6	9	4	8		1
6	1	2		5	3	9	7	4
		4		1		6		

MEDIUM - 416

			2				1	5
6	3	8	4	5	1	9	2	7
2		1			9	6	4	8
3	8	5					9	4
7	1			8	4		3	
	9	2	5		3		7	6
	7	4		3			8	9
1			8		5		6	
8		3		9		4	5	1

MEDIUM - 417

	3			4	2	5		9
				8	9		1	
8	5	9	7	1	3	6	4	2
4	1	2	3	7	6	9	5	
5	8	6			1			
	9			5		2	6	
		3				8		
9	7	5	8		4		2	6
1		8		6			7	4

MEDIUM - 418

1	7					9	3	5
		9	7			1	4	
3				2	9			
9					2	3	7	6
7	3				8	2		
4		6		7		5		
2	9		5	6				8
6	8		2	9	4		5	3
5	4	7	3	8	1	6	2	9

MEDIUM - 419

		6		1	4	2		9
7				6	9	1	5	8
	1	9		7		4	6	
1				3	5	9		4
4	2	8		9	7			
9	3	5		8	6	7	2	
6		4	9		3	5		
5	7				8		9	
2			7	5	1		4	

MEDIUM - 420

	1	5	6					
2		3	1	9	7	4		
	4		5	2	3		1	
	2	1	4		5		8	7
						3	4	1
4	9	7	3	8		5	2	6
			8	3	9	6		
6	3		2	5			7	8
8		4						3

HARD - 421

9		6			5			
4		7			2		8	1
			4				7	9
3	7		6	9			2	
	6					9	5	7
8	1		2		7		6	3
5					6			4
	4			3	1			
6			8	4	9		3	5

HARD - 422

		1		2	3		8	7
4							9	
3			9		8		2	
	9			8			4	2
		4			5	9		6
1		7	2		4	8		
2		9			1		5	4
	1			3	9		6	8
6	4		5				1	9

HARD - 423

	2	3	5	9		7		
	6	9		8	4		2	
5			2		7			9
			4	5		6		
3				7			1	
	7				3	9		5
	5				8	3	9	2
	3		9					6
	9		1		6	4	5	7

HARD - 424

8		4	7	5	1		9	
	5	2			9			4
	9		8		2			
3		6	2	7			5	
5			3					
		8		6	5		4	
		5		2				9
				9	6	3	1	2
		3	1	8	7	4	6	

HARD - 425

					5	4	1	9
5	1		7				6	
3		4	1				7	5
9			5					7
		1					2	
			6	7	8	9	3	1
8		7	2	3		1		
2		6	4	1	9	7	8	
		3				2		

HARD - 426

7			3		1	9	8	
					7		4	3
	6					7		1
1	3	5						
	7	8	2	3				4
	4	2	7			5	3	8
	2		1	7		3	9	
5		7	9		3		2	
4						8		7

HARD - 427

6	7					4		1
1	9		3				7	
4		2	7	1		3	9	5
5	2	6						9
			6	5	8	2		3
	1		2					
		4						
	6		4		5	9	3	
			8	7	3	1	4	6

HARD - 428

	4					6		
		3	2	4	1	5		
		7	5	8		3		
				1	4			9
4		8				7	5	1
1	3		9	5	7	8	6	4
3	2		6		9	1	8	
7				3	5		2	
		9						

HARD - 429

	7			1		4	9	8
4		5		7	8	1		
3		1	9	4		5	2	
	5	7	8			2		
8		6		5		3	7	9
			4	9		8	5	
2	3	9	6	8				
7								
					4		3	

HARD - 430

		2	1	7				3
	5				3	9		2
	8		9		6		5	4
		6			1	5		8
	3						4	
	2	4	7	3	8			
2	4						8	6
8	6	7	3	9	2	4		
	1				4	7	2	

HARD - 431

	9					7		5
	3	2		6		8		
	8			5		2	1	
8	1	3		2				9
	2	6	7	9				1
		9	6					8
	6					9	8	7
9	4	8	3			1	5	
			9		5	3		6

HARD - 432

		9	7					
5	8				6	3	7	2
				3	8			1
8			6	9	7	4	2	3
	3	4	8		1			
6						1	8	7
	7		1				4	
1		8	4		5			6
	5		2		9			8

HARD - 433

		2	6			4		
6		4		7		5		
	1		9	4		6	7	8
	2	5	4	3		8		
			2	5	6		4	
	6	9			7			
			7	9		2		4
	4		8	2		7	3	
2		7		6				9

HARD - 434

						3		2
		2				1	7	
3		5	1	2	7	6		
				8	1		5	4
4	1	6			9			
5	9		2	3	4	7		6
					5		9	7
	5	4	9		2		6	
8				4	6	5		

HARD - 435

2	8		4		3		7	9
4			2		9			
		1			5			
8	1		3	4			6	5
	5			7	6	4	1	
6	7			5			9	
			6		4	8		1
1			5	2	7	6	4	
			8	3			2	

HARD - 436

5	6		1		7	3		
		3	5	8	2	7	9	6
		7				5		
9	4		7		1		8	
				6				7
7					9	4		
	5						7	9
	3		9		4	8	6	
8	7	9	6	2	5			

HARD - 437

9	2		1	8			7	
3	6							2
1					6	3	4	
	3						8	9
	4		8	3	7	6	5	
8	7	6				2		
6	9	3	4	5	8			
7		2				4	9	
		5	7	2		8		

HARD - 438

	1			2			6	3
	6	8						
4	3					5	7	
1			2	6				
	4	6			9	2		
	2				4	7		6
		7	8	4	2			1
		1	7	9	5	4	3	2
	5		1			9	8	7

HARD - 439

3		6		7	2			4
		5	6					1
	1	8	3				9	6
	3			6	4	1		
			2	5		9	3	
5		7				6		2
			7	2	8		1	
8			4			2	6	5
4		3					7	9

HARD - 440

	2	8		1			9	5
	7		2	8		1	4	
			4	5		8		7
	5	2	9					
9		1	5	7			6	
				3	2		5	
6	8			9				2
		5	8	2		9		4
2	1		7	4		6		8

HARD - 441

7	5			4	1		8	
		4	5		9		2	
	9	1		6				
1					3		5	8
5		7		1		6	9	
	8	3	6	5		2		1
		9		7		4		5
6					4			9
4		8		9				2

HARD - 442

		9				2	1	6
7			9	1	2	4	8	
				5	6		7	3
3	2	7				6		
			1	2		5		
	8	5	4			7		2
				9	8	3		
	6					1		
2	4	3	6	7			5	9

HARD - 443

1	4	6	2	5	7		3	8
7	8				1		2	
	5	9						
	2				3			
			8	4	2		6	
6		7			5	8	4	
8					6		9	
5	6	4	3	7	9			
3	9						5	6

HARD - 444

2			1	9			3	8
8	3	6					7	9
	1	4			8	5	2	
7				6	1		4	
6	4			5		9		
	9			8		7	6	
1				2	4			
3	6				5	2		4
	5		7				8	

HARD - 445

						8		
2				5	4	9		
3		6	9	8				1
1	6	4			9			5
	2	3	8	6	5	1	4	
8	5	9						
	9	2	7	4		6	1	3
4								8
			2	9	1		5	

HARD - 446

		1	3					2
		6		7	5	8		9
7		4			6			5
1		7		5				8
8					4		2	
		2	9		8		6	1
	1	5		6			8	7
2		8	5	9	7		1	
6				2			5	

HARD - 447

							8	2
3	9			1	8			
4		8	7		6		1	9
	4						6	5
9			5	6	1		4	
6		1		2	7		3	
			6	7				3
	6	9	8			4	2	1
8					9		7	6

HARD - 448

8	1	4		6	7	2	3	9
				2	9			
6	9	2						
	6		2	5				
9	4	7	3			5	1	
				7	4	6	8	
	2			9	5	4		8
5	7					3		1
			6		3		2	

HARD - 449

	7		6					8
3		6		8	2		9	1
2					4		7	
1		7	8	4				
	9	2	1	6	5	8		7
		3			9	1		6
9		5	3	1				
		8	4		7			9
	1				8	3		

HARD - 450

4		8				7	1	
	1		4			2	3	5
		7		9	1		8	4
		2	7		3			8
5		3		6			2	1
9	8			2	5		7	
		5			9			2
		6		1				3
2				5		9	4	

HARD - 451

		3		8	5			1
8	6	9		4	1	3		
5	1		9			2		4
		8		2				6
		4				1		7
6		2	1					8
4			6	3				9
	9			1		7		2
7	8		5				6	3

HARD - 452

8				6				
			8				2	
	9	7				6	3	8
		2	6	7	5	9	4	1
		6		4	9	8	7	3
		9			3	2	6	
9		4			8		5	
	7			5		3	9	
1	3					4		2

HARD - 453

7	5		3		8	6		4
4				7	1		2	
	8	2		4			1	
9		8			3			6
	4			8	7	9		1
	3			6		2		
	7	9			6	5		
3				9	4		8	
	1		5	3		7		

HARD - 454

	1		4		3			9
3	5	6	9			4	2	
7	4				6	8	1	
			1	8				6
	6				5			8
5				4		3	9	1
	7				9		3	5
	9	5	8				7	
6	2	3		5				

HARD - 455

6	9		2	1				3
	4		3			1	9	
	3			4	7		5	
	5	3		6		4		
1		8			2			
7	2							
3	8				9	2		
4	1	9	5		8	3		
2	7				4	5	8	9

HARD - 456

		3		6				
6	8			1				4
	7	4	9	5		6	1	
5			6	4		7		8
		7						
	1	8	5	7			2	6
		1	3			8		
9	3		4	8		1		
	4	2	1			3	6	5

HARD - 457

	1		2		7			3
3	6	7	5			2		8
	2	4		8		1		
2		6		5				
	7		1	6		5		9
		1				8		6
			8	4			5	1
1	4	8				7	6	
			7		1		8	4

HARD - 458

5	7		2	9	6		3	8
4		2	8		5			7
			3		7		5	2
				2			7	
3		7	4				2	1
			5	7	8			3
	6	4		3	2			
1	5	3	7		4			
7				5			8	

HARD - 459

	6	4		5	7	1	9	
7	9	3	6	2	1	5	4	
	2			4			3	
		1	2		5			4
			7					9
					3		5	1
	1	8					2	
9		7		3	2	8		
				8	6	9		7

HARD - 460

2	8	6	7				3	9
		7	2		3			4
3			5		8			7
	5		6			3	9	
		1	8		4			
7			1		9			5
6		8		7		4	1	
		3		8		9		
		5	3	1	2			8

HARD - 461

2		5		7	4			
7	4	6	2		3			1
	8			6				
		4		3		1		
1					5	2	3	
					9	4	8	5
6	9	1		2			4	8
4		8				3		
5	3				8	9	6	2

HARD - 462

6	7		3	9	5	2	1	
	5	8	6		1	3	7	4
			4	8	7		6	
8		6						7
		5						
		1		3	9	6		
		2	9		6			
4	8					1	9	6
3						4	2	5

HARD - 463

		8		2	3	1	9	5
6				8		2		
					4		6	3
	1	4						2
3						4		
8		2	4			7	3	9
4	8	7		9				
		1	8		6		7	4
9	5	6	1	4	7			

HARD - 464

8		6	2		1			5
	1	4	6	8		2		3
2	7			3	4			8
3		8		5	2			1
5		2				3	8	
7			8	9				
6		3	5			4		
	2	9			8			
	5			2		8		9

HARD - 465

1	2		7		8			
8				6		2	4	1
	3			2			7	
		6			3		1	5
2	8				6		9	
	5		9	8		4		
	6		5	1	2	7		3
5				3	9		2	
	1		8		7	6		

HARD - 466

	7		9	5	8	3		4
5	2	4				8		
		9		4	2	5		
	6		5	1		4	8	2
	1		3	7		6	5	
9			2	8	6			
	9	1		6				
			8			7	3	
6	3	7		2	5			

HARD - 467

	6	8	9	7	5			4
					2	5	8	
2			6			3		
	1		2	6		8	4	
			5	9		1		
8			1		3			5
9		5	4	3	1		6	
			7	5		4		8
7		3	8	2	6	9	5	

HARD - 468

3	7	2	5	1	6			
1		5	9	8		7	3	2
			3	2	7			6
	4		6	5	2			8
	3				9			
	5					6	1	
7								1
4		9	7		1		8	
	1	6		9		2		

HARD - 469

	6	4				7	3	
	8				9	4	1	
9	1						6	
3	7	6				2		
	9	2			5		4	
		1	8		6			
	3	9		8	7			4
	2		1	6		3	9	
	4	7	9	5	3			1

HARD - 470

5			4		7	8		1
			9	3	1			
1		3		5			9	7
8		4			6	7		3
		6	1	7				8
7		1	3				6	
3	7			1	5			
6			7	9		1	8	
9		2	6		8			5

HARD - 471

	5				6	3	8	9
3	9		8		1			2
	4		9		2	6		5
	6			1	5		2	3
8	7		2					4
5		3				1		
			3		7		4	1
6	1		5			2		8
4							5	

HARD - 472

					4			
1			9	6	7			2
8	9		2	5			6	
7	2						4	1
9	3	4	5	1	2	7		
	5		4					9
5	7	9				3		8
	1					6	9	7
		6	7	2		4		

HARD - 473

2	9				5	3		
		8	9	3		2	6	4
		3		4	2	7		9
	1		2	5		8	4	
	4	2	3	9			7	5
		5					2	
4	2	9				5		
			7	2			3	
	3	1	5					

HARD - 474

	9		2		4	1		6
	1		6		5		7	4
	5	4	7	1		9		8
7	4		1	2	8	5		9
2	8		4					
						4	8	
				6				1
	2			5		7	6	
1	6	8	3					

HARD - 475

					6			3
			7		3			
6	3	8	5	2		1		9
	2		8		9	3		5
8		3			5	9	1	7
	1				7	2		
3		2			1	7	9	4
	6		4	7		5	3	
	7	5				8		1

HARD - 476

6	1			5	7	2		3
3	7		9		2			6
	2		6			8	7	9
1	5						3	4
9	3			1				2
	8	2				6	5	
8				7	3			5
2	9					1	6	
	4					3		

HARD - 477

			5		8		4	7
7	8	3						
1			6	2			9	3
		7		4		6		
8	9	5			6	2		
4		6			3			9
	4			6	1		7	5
5	3			7	9		2	6
6	7							1

HARD - 478

	4					8		
7	1	8		4	6			9
2	9		3			4		7
	5	9	4		8			6
	3		7	6		1		
	6	7		2		5	8	
3		6		1	4			5
9						7	4	8
				7			3	

HARD - 479

		1	4		5	2		6
		7	1		3			
	3	4	8	9	6	7		5
9	8	6		3	1	5	2	
7								1
					9			7
		2					5	
	6	9			4	1		2
1		8	5	6			4	

HARD - 480

			2			1		
		1		8	9			4
6		2	4		7	3		
		5	3	6		8		
1		3	9			5	4	
7		4						3
	2	6	1	4		7	3	
						6	8	5
5			8	9	6	4		1

HARD - 481

	4	9	7	2	5	6	8	
			9	1		2	4	
	6						9	1
		1					2	
6	3	4		9				
				6	1	9		7
		6			2	3		
2	1	3		7		8		4
	7			4	3	1	5	

HARD - 482

5	7			4	6	3		1
9	1		2		7			
4	3		9		5			
7			5	8		9		2
8		9			4			
6		3			2		4	8
							1	5
2	6		3					9
	9			7		4	2	3

HARD - 483

1	8	7	6	9		4		
	2	6	7		4			
9			1		2		6	8
	1				9	5		6
6						2	7	
3		5	2	6	1	9		4
8	9		5	2			4	
4		2						3
					8			

HARD - 484

	1	8		6	5			
3	5	4	1		7			
				3		1	4	
		2		7	1		5	
		1		5	3	7		
	3			4			8	
	4					8	7	6
6	8	9				5	2	
2			3	8	6	4	1	

HARD - 485

7	5	2	1	6		8		4
6					8	5		3
		3	4					
				1				
8			2			9	4	5
				4		1	8	2
3	6	8		9		4		1
	4		5	8		6		9
1				3		7		8

HARD - 486

			5		4	8	3	1
			6					
2	8		7				5	4
1	2	4						
6	5	7			3			9
	3			6	7	4		
3		8			1	9	7	
	4	2	3	9		5	1	8
				7	2	3		

HARD - 487

1		6	5	3	8	9		2
	8		4	6	9	3		1
9		3	1		2	6	8	
7					3	2		
3	2			4		8		
		4	2		1		6	
6		9	3					8
								7
4	3			8				

HARD - 488

9	3		2					5
		2					7	1
7	1		6	3			2	4
	6							
3				2	9		1	6
		9	7	1		2		
1		3	9	6	2	5	4	7
4	2	6		5				8
5					4	6	3	

HARD - 489

7		1		9				8
	5	9	4	3			7	
		8		6		2		
3	1			4	5	8		9
				7			2	
	2		3	8	6	5		7
1	4		7	5				
5	7					9	3	1
				1		7	5	

HARD - 490

	1		7		3		8	
	8			1		3		
	4	3			2		7	5
		9		5	7	2		
	3	7		2	8		5	4
	2		9				6	3
		1	4		5		9	
	9	8		7			4	
2	5		1				3	

HARD - 491

	8			2	4			6
4					6	2		5
				7	9	4		3
	3	6		4		5	7	
8						6		1
	2				1		4	9
	5	2		8		1		
6			2	9				
3	4	7	1		5		2	8

HARD - 492

9	6		8	7				4
7		8		4			2	9
			9			7		8
4		2		6				7
	9		3	2				5
		6				3	4	
6	8		1		9			3
	7	5				4		1
	2	9	4	3		8		

HARD - 493

		9	6				4	3
	1	3					6	8
	8			3	9			7
9		5				6		
6		2				8	9	1
1			7	9	6	3		
8			2		5			6
3		4		7			1	2
5		7	3	6	1			9

HARD - 494

		7	1	6	4		5	
5	1		7	2				3
2		8	5		3			1
3			8	1			7	
	8			5	6		3	
6								9
			6	8		3	1	
1	6	3			5			
8		9	3				4	6

HARD - 495

2	8	7	4				5	
9	4		1		5	6	7	2
	6		2				9	4
	3	5		2	4		8	6
			5	7				1
			3			9	4	5
	7		8			4		
	5				9		3	
	1	9			2		6	

HARD - 496

		5	8	1	7		4	
3		8		4	6			5
2		4	3		5			1
4			5	7	1			6
9	5	2		3	4			
7				8			5	
			7	6		1	2	8
	2					7		
1		7	4				6	9

HARD - 497

	6		8					
1	5	7	2	3				6
		8			7		1	2
9		2	7	6	3		8	5
8					9	2	6	
5	1	6	4		8			
			9	7	2	4		1
					1			9
					6	7		8

HARD - 498

	3	7	9		2			
	2				6	1	3	4
	6		3				2	9
	1	4	2		7	5		
					4	9		7
		3		5	9		4	6
		6	4		8			
	7	8	6				9	
	4			9	3		6	2

HARD - 499

1	5	3	2		6	8		
2								
6	7	4			5		2	
9	2	1				4	3	
		5	1		4	9	8	
			6	2				
			9					8
8		6	4		3	2	9	
3	9			8	1		4	7

HARD - 500

	1	2	6				7	4
			1	9	2	8		
		3		7				
1				4	7	3		
6	7			8			5	2
3	2	4		6		1		
	3	6		2			1	5
	4	8						
2	9	1	7	5		6		

HARD - 501

			5					
	1	3	8		7	6		
			2	6	9		8	1
2		8					4	6
7	4						3	9
	9	6		8		2	5	7
1	3	7	4	5		9	6	8
				7	6	4		
	2			1				

HARD - 502

5	6			3			4	1
		2		8	1		3	7
1		3	7	4			9	
4			8	5			6	
	2	5			4			
	3	8		2		5		
3		4					2	
		9		6		1	8	
8	1		2		3	4		

HARD - 503

7	9			8				2
6	2	4		3	7	8		1
	5					7	3	
			9			1		
4					8			
	3		2	1	5		6	
	8		7		6	3	4	9
3	4		8	9				
9	6	5		4	3			

HARD - 504

	1	7	6				2	
2	8	3		4	5		6	7
4	5		7					
			9			7	1	
	9	1		7	3		4	
7		2		1				
1		9				3	8	
3	7	4				2	5	
5	6		3		9			

HARD - 505

			2	8				4
	6				5			9
				3	6	1	8	7
	2				3	8		1
	9	3	7					
1	8	4		2	9			
6	4		3	5				2
2	1	7	8				3	5
	3				2	4	7	

HARD - 506

3		4	1	9				8
		1		8	5			
			4			7	9	
8	3		5		9		1	
4	1		8					3
				1		9	8	6
9	8		7	4	3			5
6	4		2		1	8		9
			9	6	8		2	

HARD - 507

1	7		4		3	5	2	6
		9	8	6	5		7	3
	6	5		1	7			9
		3				2		7
7				3		9		5
9		6			2	4	3	
4							1	8
						3	9	4
		1			9	6		

HARD - 508

8			7		1			2
2	7		5		9	1	4	
			4	2	8			
	1		2	5		9	8	
4			9		7			
			6		3			4
5	8			9		3		7
3	9	6		7			2	
	2			4	6		5	

HARD - 509

	8				2		1	5
4		6			1	9	2	
2		1	8		5		4	
	6	9					5	
1						4		9
5			9	7		1		
6	9		1			3	7	
7		8				5	6	
		4	5		7	8		1

HARD - 510

4		7	2	1			8	
3	2	1		9				
				3	7	4		2
				2		1	5	
	1	2	7	5				3
5			8	6		2	9	7
2	3				5			
	8	9				7	3	
6		5		8				4

HARD - 511

			6					
				2				5
6			5		8	4	3	2
2		7			9		8	1
8	4	9	1					
5					7			
3	5	4	9	1	6	2		
1	9	6			2		5	4
			4	5		9	1	6

HARD - 512

	3	5	7		8		2	6
		9		5		7	1	
	8	7	6	2				4
	1		2					
		3	8	6		4	5	2
6	5	2	3	9			8	7
3	4	1	9	8		2	7	
					7		4	
		6		4	2			

HARD - 513

6			9	2			8	
2	7		8	4	3	9		5
		5	6	7	1		3	2
				9		7	4	
	6			1		5		
	1		4	8	7			3
			1	6	9		5	
1	9		2		8	3		
			7		4			9

HARD - 514

							8	2
6	2		5		8	9		4
8	5				4		6	7
9			3	4	6		2	
	7	5						3
	3		2					
5			8	1	3	4	7	
		7		6	2			1
1	4		7		5			6

HARD - 515

	4			8	7		5	9
		7			5			
		8	4	9	1	7	6	2
	8			5		6		
					6	4	8	3
4	1						9	
8		4	7			2	3	5
5	7	9	2	4				
		2	5		8			

HARD - 516

	2		1	9				
	4			5	8	6		
5		8		3		2	9	
	9	1			5			3
7		2					4	9
3	8	6				7		
		7		1		3	8	
	3	5				9		4
2	1			8	9	5		6

HARD - 517

7			9	2				4
				4	6	7		3
6	5		8		3	1		
	2	7	3	8		6		
	1			6	7	8		2
	6	3		1	9	4		7
	8		1	3		9		
3				5				6
		1	6				4	8

HARD - 518

4	7					5		2
2	5	6	1	7	3	4		9
	1					3	7	
		7	9	4	5			
6				8		2	4	
1					7	9		3
						7	3	
		8		9			2	4
7	4	1		3	2			

HARD - 519

				9		5	4	3
8		5		4	1		2	
9	4	2	6	5	3	8	1	7
		7	8	6		3		1
	9						5	2
2								
	8		4	3	6			5
	7	6	9	2		1		
						6	3	9

HARD - 520

9	6			8	7	5	2	
2		1		9	6	8	3	
	7		5	2		6	9	4
	8					2		
1		6	2				5	
			6	4	8		1	3
6		8	9					
3	9		7					
		2			5	9		

HARD - 521

			1		5		3	
7	8	6			4			5
		5			9	4	6	2
2		4		7		3	9	1
				5				
1		8		3				7
		9	5					
5	6	3		1	7	8		9
		1	6	9		5	2	

HARD - 522

8		1		6			4	2
2			1	4	3			7
								6
		9		3			6	4
	2	6		8				1
		4	6	7				
1		8				7	3	9
			7	5	9	4		
9		7	3		8	6	2	5

HARD - 523

		3	5		4			
2	4		1			9	5	6
	8	9	7		2	1	4	
		2	6				8	7
3		8			9		1	
	1	5		4	7	2		9
				7			6	
8	3				5		7	
7	5	1	4		6			

HARD - 524

8					3			2
		7	1		6		4	
4			2		5	6		
	3	5			9	7		4
			5		1	2	9	
	8	9	7	6		1	5	
	6				2	9		7
		2	6			8		5
7	5		3			4		

HARD - 525

	5				8	9		4
9								6
		6	4		1	7	3	5
7	2	9	5	3				1
1		5		8				
		8			2	5	4	9
		3	2	6	5			8
	9		8	4	7		6	3
	6							

HARD - 526

		2		9	8	5	1	
7	1	8		6				2
6						8	7	
			2					9
8	2	4	3	7			6	
	7	1	5	4	6			3
	9		8	3	7	4	5	
			6	5				
		5	9	1	2	6		7

HARD - 527

8	3	4	1				9	
	7			4	2		8	
	2	6		9	8		4	5
	5		4				1	3
4			5	2			7	
		9		1			2	
	9					7	6	1
			8	3		4		9
		1		7		8		2

HARD - 528

	5	1		6			8	
	8	9	2	1			7	4
		2	8					
	6					3	1	8
	3		9	8				
5		8	6		3	7	2	
1		6			8		9	5
7			1		6			3
		3	4				6	7

HARD - 529

			9	1	7	3		
	8			2		7		
			8					
	9	2			3		5	4
6			4	8	2			9
	1	8			9		3	6
1	6						8	3
	7	3		9		4	6	2
2	5	4			8	9		7

HARD - 530

	7			2	1	6		5
	6	3			8	9	1	2
	9				3	4		
	4			3		2	8	1
		2	8	9	5		4	
8	3							6
						1		4
	2		1					
1	8	9	5	4			6	7

HARD - 531

	7							6
9	8	2		3				7
3	5	6				8	2	4
5		8	4	7	3			1
	9		8			7		
	4	7	1		5		8	
8		5			7	6		
2			3				7	
7				2	1	4		

HARD - 532

	6					1		
4	7	2		3	1			9
5	9	1	6			3	4	8
	3	7	5					1
	1	5	7	2				
6	4	8		1		2	7	
7		9	3	8	5			4
3						9	5	2
				9	4		8	

HARD - 533

	1	7		3	2		9	
2				8	6	7	3	5
		6	5	7	9	4	1	
9	2			4	1		7	
	3	8			7		6	
7	6				3			
1		9	2		8		4	
	4				5		2	
		2			4			9

HARD - 534

9			2		5	6	7	3
3		1			9		5	8
5	2	6	7			1		
4			1		6			
2	8	5		9		4		
1		9					8	
				6	1	5		
		2	9				3	
8				2	3		6	9

HARD - 535

		4	5				9	6
	8	9	3		1	2		
	5	1		6	9		8	4
		6			4	5	7	9
	9				6	8		
4		8		5		6		
		7		1			3	2
	6			9	2			
9		2	4		5			8

HARD - 536

8			5		6	4	1	
		4				6	8	
	6						3	9
			8		2		4	3
9		1			5	2		
4	3		7		9	8		
3		8		9			7	
1	2		6			3	9	
6	5	9	3					8

HARD - 537

		1	9	8			3	6
5	3	8	1		6	7		9
	6	2		3	5		4	
	8	4	6				5	3
6			2			4		
		3					7	
3		7		2	8		6	4
					9	2		
	2			6	1			

HARD - 538

		6	4					
		3	2		1	7	6	
	7		5	8			2	
6			7					4
	3	9			5	6		
7	4	2		1		9		5
3	6				2			7
		4	1		8	2		6
2	1	7		6	4		5	9

HARD - 539

1	6	7	2				9	
8			1		6	3		2
			8	4			7	
	5					9	4	3
6	9				3		8	1
3	2		9		4			7
	8		3		7			5
9	1			6			3	
4		3	5		1			

HARD - 540

	1	3	4	9		7	6	
4		8			7	9		3
2		7		3		8		1
8					3			
3	4			1	9	6		8
		1	7				2	4
1				6	2		3	
7				5	4			
	2	4				5	8	

HARD - 541

3	7	8			1		4	
6						8		
	9		4	8	7		1	
			5		3	7		
	3			6				
2	6		8	9			3	1
4	8	6	1		5		9	3
7				3	8	6	5	4
9	5	3		4				

HARD - 542

		2		4				9
4					6		5	2
8		5	1				4	
	7	8	9	6		5		4
		9	3		4	1		8
6					2	9		
1		6		9	7			3
				5			9	1
	4		6		8		7	5

HARD - 543

	3	1			4	7		
6			9	7	8	1		
				6	3	5	2	9
4	6	3	7	5		8	9	1
5		7				6		
			6		9	4	5	7
					7			5
	1	9	4					
7	5					3		

HARD - 544

		9	8	6				
8	1	3						9
	5	6	2	9				3
		4	6			3		2
			9		5	6		8
5		8	1			4		7
			4		9		7	5
7	8	1			6			
	4	5		8		1		6

HARD - 545

5						1	4	8
	3							2
			8				6	5
8	9		5	2				
	5	4	6	1	9			
6	1		3	4	8		5	9
			2			4		7
3		2		8	1		9	
	6	9		5	3			1

HARD - 546

							9	
3	8			9		2	6	1
			7		2	5		4
	3	5	6	2		1		
	1			7	9			
			1			6	4	
6	9				1	7	3	
7			9				1	8
8	5	1		4	7		2	6

HARD - 547

	2						9	5
7	6		9		5	1	2	4
						7		
3		6		9				
			6	2		4		
	9	4	5			6	3	
		1			9	2	7	6
9					4	5	8	
6	5	2	8	7			4	3

HARD - 548

		3	9	6		8		5
6			4		5	9		7
					8	3	1	6
	9	7		5			6	
8	2			1			9	3
	4		6	9		1	7	
		1	5			2		9
5			1	8			3	
		4					5	

HARD - 549

	2	5		7	9			8
	9		5	2		6		
6	7	4	8	1			5	
9	4					8		
		1			7	4	6	
7		8	3				9	
3	1					5		
5	6			9	1		8	
		7		3		9		6

HARD - 550

4		8			9	6		
7			6	3		8	2	9
		6	7				3	5
8	7				5	9		2
2	4		9	7		3	6	
1		9						7
3		7	1	9	6	2		4
6				8	2		9	
		2						

HARD - 551

			4				8	
		1	5		8	7	2	
			1		3	5	6	9
1	7	5			6			2
	2						1	6
		9		1		8	5	7
6	1	7	2	8			3	5
					1	6		8
8	9			7				

HARD - 552

		4					9	6
3	2		8			5	4	1
1				4	9		7	3
	8	5	9	1				
	3							2
			3			4		
5		2		6	8			
	7		4		3	6	2	9
6	4		7	9			5	8

HARD - 553

9		8			2	3	6	
			9				5	
	6		3					
8		5			4			3
4			2			5		
3	7	6		8			2	
1	5		8			7	4	2
6		7	5	2		9		8
	8		7		3	6	1	

HARD - 554

	2		7		9	3		5
	6		2	3			1	7
	7	8				2	9	
					7	6	5	
6		5	9					
	3	7			4			9
	1		4	9		5	3	
		6	8	1	3			2
	8	3	5				6	

HARD - 555

5	7					9		6
	2	9	4			8		
8		1	9	2		3		
		2				1		3
	8		5		1			2
		5					9	8
7	1		2			5	8	9
				9	8	2	1	
2	9		7	1	5			

HARD - 556

7		3		4	9		5	6
1	4	6		8			7	9
	2	5				4	8	
	7			3	6			
2	6		8	5	7			
	5		9			8	6	
					4	7		
6	9		1	7				4
	1		5		3		9	8

HARD - 557

	9	1		7		4	2	5
				1			9	
4	3		9	2	6	8	7	
		9	8		5	2		
2	5						8	3
	1						4	9
1	2		6	5				
	4		1		7	9		2
			4	3		1		

HARD - 558

							4	
8	3	4				1		7
	1	5					3	6
1		7		4	9	6		
		9	2	1	3			
3	4	2	7	6	5	9	1	8
	7	1	6	9	4	3		2
4	9			7		5	6	
					8	4		9

HARD - 559

1	5	9	4					
3	2	4		6		7		5
	8	7		2			1	9
			2	4	6			
			7			1	2	
2				5	3			
8				9		2	6	7
9		2					4	
		5	6	1	2	8	9	

HARD - 560

			6	5	4		3	
6		9			8			2
					2	6		
	9		1	7		8	2	
2	7		5				9	1
	3				9	5		6
1	8			3		2		9
3			9		1			7
9			2	8	7		5	3

HARD - 561

1	4				7			
		9	1	5	2		7	6
5	7			6	4		9	
6	5		8		1			
		1	6	7	3			5
7			2	4				
	1	7		2			8	
		8	5	1				7
		5	7	3			2	

HARD - 562

6		8		3				9
9				5		2		3
		3	4		7	6	8	5
5	2		3			8		7
3	6		7	8				
		4	1	6		9		2
1		6		2	4			8
8			9	7		5		
			8		3	7		6

HARD - 563

6		2			3	1	8	9
1			2	8			4	
8			9		1			2
9		8		3			5	7
	1	3		5				8
		4	6				2	
	2	5					1	3
4				2	7	5		
3		6				2	7	

HARD - 564

9	5	8			2	3	7	6
	2					5	1	
		1	5	6				
					4			
			6	2	1			8
6						4	3	7
2			7	1			4	3
		4	2	8	3	7	9	
8	3			5	9			1

HARD - 565

5	2			3				9
						8	7	4
7			4		9			
		1	2	4				
3	5	2	7		8		4	
4	6	7		9	3	2	5	8
2		3		7			8	6
				5	1	4		
	4				2		3	

HARD - 566

	7			2				3
	2	6				8	7	
	3		7		8	6	9	
7		4	1				6	
2	1	9			6	4		
	8	3		9	7	5	2	
			2	8	4			6
	6	2	9				4	
	4		3			2		

HARD - 567

8				2			1	
					1	8	7	6
1		5		9	8	2	4	
2	1	8	7		6		3	9
		3	1	8		6		7
5	6						8	1
		2			3			
6	4	1	8		9			
	5	9						

HARD - 568

		5			3	4		1
		1	9	4				
3		2	6		7			
	2			6	9	3		5
9	7		5				4	6
		3		8	1			2
5					4		1	
2		6		5			9	
	3	7	1	9		5		8

HARD - 569

	3					8	6	7
6		8		3	2		5	9
			5			2		
1	9			7				
		6	4		8			
8	5	3	6		9		4	
	8		1	6	3	5	7	2
2					4		9	
3		5	2				8	

HARD - 570

3			7	1		2	8	9
	2		4	3	8	5		6
	1	6	5		9		7	3
4					1	6		
	5			6		9	3	4
			3		5	8		
1	6		9					5
		4						8
	7			5		3		

HARD - 571

	5		3	1			2	6
	4		2			3	8	
			9		7	1		
4	6		8		1			3
	1	3		9		8	7	4
	8		7	4		2		
6	9			7				8
	3	2	4	5			1	
		4	6			9		2

HARD - 572

	7		1			6		9
3	4	6						7
8		9	7		5		2	3
4	8	3	2	1	7			6
7						8	3	
		2	5	8				
			6		9			8
			3		2		7	4
				7	1	3		5

HARD - 573

	4	1		7		6	5	
7		6				3		2
				4	5			
	9	3		6			1	4
6	1			3		8		7
8	2			1	4			
	6	8				9	3	
2		9			1	4		6
	3	5		9		7		

HARD - 574

		3		2	4			6
	7	4	1		6			
6	5	9				2	1	
			2	8	3			7
8						4	3	1
				1	7		9	
	1					3		
7	3	2		9	5			8
9	4	8		3		6	2	5

HARD - 575

				8	1			
							9	8
6	3		9			7		
5	1		3				8	4
2	4			5	8	9		
			4		6	3	1	5
9	8	7	5	1		4		
		4	7	3	9	8	5	2
3	2			6				

HARD - 576

5				6				
9	3		7					2
2		1	9		3	7		
	5		6	1	7		2	
1	9		5					7
	7			9	8		6	
	8			2		1	4	6
6	1		8		4			5
4			1	5		9		

HARD - 577

6	8	7			4			
	3			8	6	2	7	4
		9			1		6	
9		2		6			5	7
	6					1	2	
7		4		1	9		8	6
			1	4		7		3
3	9	1		7				8
			3				1	

HARD - 578

2	6	1			8			
		7	2	1	9	3	6	
	9		5	7	6			
7		2		5			9	3
						7		8
		4		2	7	6	5	1
			7	3		8		5
1	7	5	9	8				6
3								

HARD - 579

		2	3			5	7	6
		9	2	6	4	1		
8	6	3			1	4	2	9
			6	2	5	3		
	5	6	4	7	9		1	8
		7	1				5	
6			8			7		1
		1						
7		8		1			4	2

HARD - 580

	2		5	4	3		9	
	7			9				1
	9		8			3	2	4
6			9	3		2		7
		7		2		6	1	8
2	8					4		
			3			8		5
	5	9	2	8		1		
8	6		1				4	

HARD - 581

1			5	4			7	
8		4		6	7			1
	6	5	1	2			9	4
						7	6	
		1	4	7		9		
6		2	8		5			
		9		5	2		1	
	1	6	9		4			
4		7	6	3	1			

HARD - 582

6		5	8			7	4	9
7		9		6				
			7			6	1	2
8			6		2		7	5
2	4		3					
1				8	9	2	6	
5		2		4				
				7	3	8	2	4
3			9	2	6		5	

HARD - 583

4		9		2	1		6	5
2		1		6			3	
8			3		4			2
	4	5		3				6
9			4	1	6		5	
6		3		5				8
	9		6	7			2	
5				4	9			3
		4			5	9		

HARD - 584

4	7			3				5
1		8	4		2		9	7
	5	6	9		1			4
		4				7		2
	8			6	4	9		
	1	2		9		4	6	8
8					6	5	4	
		5		8				9
	2	3		4		8		

HARD - 585

				7	3		6	2
			1	6		9	3	
6	9		5		8			
3	1		7		6		2	
9			2	8		7		3
7	2		4	3	5	6	1	9
	6		3	4	9			1
						3	7	
4	3							

HARD - 586

	5		7			2	3	1
		8	5		1		4	
6	1		9		4		7	5
2						3	5	
		1			3	7		
		5	6					9
	7	6	2		5		1	
	4	2	3		6	5		
5		9	1				2	

HARD - 587

2	3		1			8		
	7	6		2				1
1		8	5		6		7	4
7	6		8	4	1		9	2
8					9	6	1	
	1	3		6				
				8				
		9		5				7
6				1	4	5	8	3

HARD - 588

5		3	2	8	4			9
	7				5	3		2
		2	7			8	5	
1				7	6			
7			5	2	8		6	1
6	8				3	4	2	7
3				4				
2		7	8		9			
4						7		5

HARD - 589

6	5				1			
					4	9	8	
9						1		6
		6			3	8	9	7
5	8	3		9				
	9	1		4		5		3
1				8	9	7	6	4
	6	9	1	3				
		2	4	6	5	3		9

HARD - 590

		6				7		
	8		7			2		
7	9	1		4		8		
	3			1		5		7
		2	3			4		6
4	6		8	5	2	1		
3				8	6	9		
	2	8				6	4	
6	7			2	1	3	5	

HARD - 591

		8				6	5	
2			6					8
	6	3		7			2	
		2	7	5	4			
	1		3		2	9	4	
5			9			2		
7	5		4	3	6	8		2
		6	1	9		3	7	
	3		8				6	9

HARD - 592

		9	3	4				
			8		5		9	2
			2		9	4	7	8
	2					8	6	4
	3	6	9		4	1	2	
8								7
4	7	3	1	2			8	
	9	8				2	1	
		2	5	9	8	7	4	

HARD - 593

5				8	3		7	
1	6							
2	8		1	5				6
6		2	9				1	5
9	5	8			6	3	2	4
4				3	5	8		
		4	3				8	
8				7			9	
			8	9	2	6	4	

HARD - 594

	1	8			7	2		9
6	9				2			
3		2		4	6			8
2	8	5	1		9		3	
1	6		5					7
7		9	6	3				
	3	6				9		4
9	2							3
8			2	9	3	6		1

HARD - 595

3		4	1	9			5	
7					5	2	3	4
8		5		3	4			
						1	8	6
	8		9		1			3
1	3		8					2
	4				9	3	1	7
5	1			7	2		6	9
6		9	3					5

HARD - 596

		2			6	4	5	3
	8			5	4	9		2
	4						6	
		7			5	2	8	
	9							6
8	5	6		9		1		
	2	4		3		6	9	
	7		5	6	2			1
3		5	1		9			8

HARD - 597

4	8	5					9	
3	9		1	4	2		7	8
	7	1				4	3	6
		8	3					
			4		9	8		
		4		8				1
8					5	6	1	4
	1		6				8	
6	4	3	8	1	7			

HARD - 598

5				9	4	6	2	3
	3	4			2		7	
	9		3		6	8	5	
9	2	1	4					
	5				7	3	1	
3			5	6				
1	4	9	2		8			
	6	5		3				
		3		4		2		1

HARD - 599

9		8	6	2		1		
		3			8			2
	4		5		1			9
4		9					1	6
6	2		3			9		5
7		5	9	1	6	4	2	8
1	9			4	5			
3		7				2		1
						5		

HARD - 600

2	4	3				9		7
9				7	8		4	2
	7		4	9	2			
					4	7	1	
	5				1	2	3	9
1	8		2					5
	2		8		6			
	1	4		5				
6	3	8	1	2	7			

HARD - 601

2			1	8		3	6	
6	5		7					
	9	3	4	6	5			7
	8	5					7	
			5	7	8	4		1
			6			5	8	
5				1	4	9		2
4				5		6	1	8
8		1					5	

HARD - 602

		4	5	2				6
	7		8	4	9	5		2
				6	7	3		9
7		9	3	5		4		
5				1	4			7
		2		7	8	9		3
	4		7					5
8		7		3	1		9	4
6	2				5		3	

HARD - 603

	4		1		7			
	6				8			2
		9	4	6	2	7		3
			3	4			7	
8	3	1					5	
				8		3		6
7	1		9	2	6	8	4	5
	8		7	5	3	9		
		2					3	7

HARD - 604

		5						
8	2	3			4		6	
				2	6	5		1
		1		5	3	4	9	
5	9	2	7	4		6		3
	3			9	2	7		
1		9	2	3			4	6
			8	7		1		
3	5			6				

HARD - 605

	5	7			9	2		
		9		5	8		1	
8		1		7	4		9	
	7			2	5	9		1
5		2	4	8	1	7		6
6			7			4		
9							7	4
		6	9		7			5
7		4				1		

HARD - 606

7				1	9	2	6	8
		9	3	8	2			
			7		6	5	3	9
5	1					7		
	3		4		7	1	5	
4	9		2		1		8	3
			6				2	
2	6		1			9		5
	7	1						

HARD - 607

			1	4		2	8	9
1	9		6		2			
2			5		9			3
	8				1		3	2
7			3	2	8			
9	3		4				1	
			7	6	3		2	5
5	2		8					1
8	6				5	3		

HARD - 608

	9				8			
6		1				2		
		4		7		5	3	
				8	2			7
3	5	7		4	6			2
			5			4	9	6
9			2		4	1		3
1	3	2			9		8	4
		6			3	9	2	5

HARD - 609

					4	5	3	
			1	3	2		8	7
3					5		4	
	4							6
1	7			5		2	9	8
6	5	9	2	1			7	3
9		5	8		3			4
8	1	2		4				
			9	6			2	

HARD - 610

		9		5		8	2	
4		2				6		
6	5	8			1	7		3
	2			4		1		
	9	7		3				2
8	4			2			7	5
	6	4			5			
		5				2	1	4
9	3	1		8	2			7

HARD - 611

8	5	6						
4	9	1	5	2			7	
3	2	7	9	1	8	5		
							3	9
	7		2	3	9			4
9	1		6				5	
	6				2	4		5
					5		6	2
	8	5	4				9	

HARD - 612

				1		6	3	
		6	3		4		2	
				5		9		
		9	5	4			7	
	7	8				4	9	3
			7				6	5
	6		8	3	7	1		9
9		1		6				2
8	5			9	2	3	4	6

HARD - 613

2	9	6	7		3	4		
			6		9		8	2
7		8		5	4	3	6	
9	2	1	5		6			
8		4				1	2	6
					1	5	9	
		7			8		4	5
5		9	3			6	1	8
1					5	9	7	

HARD - 614

5	4	6				7	9	
7	2		4			5	6	8
9	8		6		7			4
1	3	7			4			9
4			3	7	8		1	
	6						4	
		4	2	8			7	
					5	4		
3		8	1	4				2

HARD - 615

	2	5			3			9
7			2	6	9		1	
			7	5		3		
9		4		2	1		3	5
1	5	7	9	3	4			
3	8					4		1
2			5		6	9		
	4		3					2
5	7			4				

HARD - 616

7	9			3	5	2	4	
				4	2		3	
3			8		6	5		
9	5		2			1		
				6	4	7		
6	7		9	8				2
1			3		7	8	2	
		5			9		1	7
2	3	7		1			6	5

HARD - 617

3	9	6				1	7	
		2	6			5		
4			2			8	9	6
	3	5	1			2	6	4
		4	3		6		1	8
1			7		4			9
6						4		
	1	7	9		3	6		
5		3		6				

HARD - 618

5			6					9
			9		1		7	6
	8			7				3
	6	2		9		7		5
	5		4		6		3	
	1	3	7		5	6	8	4
2			8			4		
		7		6	2	3		
8		6			9	5		7

HARD - 619

9					7		1	3
4		1					2	
		6	3	1				9
	4	5				3		
1	9	3	5	4	6			
2	6	7	1	3			9	4
		9				6	4	7
3				5	9			
6		2				9		5

HARD - 620

5	3	8		6			9	1
		4	5			8		6
7	2		8				5	3
	5		1	7		3	6	
			4		3			
9	8		6		5			7
	6	9		1			8	4
		7	9	5				
8	1			4				9

HARD - 621

7			1			4		
3	9		7					6
	5	6			3			
	7			4	1	3		8
9		1		3	8	6		
5		3	6	7		2	4	1
	6	7		1		9		
				2		1	6	5
	1		9				8	

HARD - 622

	2	3		5		4		
					9	5		
		7	4	3		6		
	4		9					2
2	6	9				7	4	5
5			2	4	7	9		6
3	7	5			1	2	9	4
				9			6	7
8	9	6					5	

HARD - 623

8		6		3			7	9
	1			2		8	5	
7		2	8	5	6		3	
				9	2			4
5	4	9	3		1			
2	6		4		5			
				6	8	9	4	
		4						
9	2	8				1	6	7

HARD - 624

9	5		4				6	
3	4				9		7	
	7			6		3		
8	1	9			6		2	
					3			4
	3		7		2	1		6
	6		2	9		4	8	5
		4	3			6		7
			6		4	9	3	2

HARD - 625

		2	7		4	6	3	9
	7		5					4
	4		6		1			2
		5	4		2	3	6	8
	1						2	
	8			5	7		4	1
	3	7	1				8	6
		4					9	3
			8	3	6			5

HARD - 626

			5					
4	7	5		9		8	6	
	9				6	7		
5		6	7	2		4		
	4	3			9			
			4				7	
	8				5	3	1	6
9					2	5	8	7
3	5	1	6	8	7	9		4

HARD - 627

	9	3	8		7		6	
			3	9				
	5	7	6	2				4
	1	2			3			
	7		2			1	5	6
	8						2	3
9	6				2	3	1	
		1		7	6		4	5
	4	8		3	5			2

HARD - 628

	4		7				6	9
8			1		3	5	2	
		5		8	6			1
	8			9	7		5	3
	9	4	6		8	1		
	7	3			1			
	3		8	6				5
	5	8	3			7		6
	1	6		7	2		8	

HARD - 629

9				6	5			
	8	4	2			6		
6	3		9	8	4	5		
	9	3			8	7		4
	5	8		4				
4	6		7	9	1	3		
	4				7	1		5
					9	2		7
2				1	6			9

HARD - 630

2			3	9		7	5	
5		8			2	3	1	
	4						6	8
			5		9		7	3
7		3			6	5		
9	1		8		7		2	
					1		4	7
6	5			8		1		
	9	4	7			6		5

SOLUTIONS

To view and download the solutions for this book, please visit **https://brainybird.org/solutions2**

You can also view and download the solution by scanning the QR Code below.

If you have any questions or feedback, please contact us at **https://brainybird.org/contact-us**

Thank you!

Made in the USA
Las Vegas, NV
28 April 2024

89239650R00066